Peter Nick (Ed.) • *Plant Microtubules*

Potential for Biotechnology

Springer

Berlin
Heidelberg
New York
Barcelona
Hong Kong
London
Milan
Paris
Singapur
Tokyo

Peter Nick (Ed.)

Plant Microtubules
Potential for Biotechnology

With 38 Figures

Springer

Dr. Peter Nick
Institute for Biology II
University of Freiburg
Schänzlestraße 1
79104 Freiburg
Germany

QK
725
.P5655
2000

ISBN 3-540-67105-6 Springer-Verlag Berlin Heidelberg New York

Springer-Verlag is a company in the BertelsmannSpringer publishing group.
© Springer-Verlag Berlin Heidelberg 2000
Printed in Germany

Cover design: E. Kirchner, Heidelberg
Typesetting: Camera ready by P. Nick
SPIN 10707141   31/3136 - 5 4 3 2 1 0 - Printed on acid-free paper

Preface

Manipulation of plant architecture is regarded as a new and promising issue in plant biotechnology. Given the important role of the cytoskeleton during plant growth and development, microtubules provide an important target for biotechnological applications aiming to change plant architecture. The scope of this book is to introduce some microtubule-mediated key processes that are important for plant life and amenable to manipulation by either genetic, pharmacological or ecophysiological rationales. The first part of the book deals with the role of microtubules for plant morphogenesis. Microtubules control plant shape at three levels:

1. Control of cell expansion: cortical microtubules define the orientation of newly synthetized cellulose microfibrils and thus the mechanical anisotropy of the cell wall. Transverse microtubules are a prerequisite for stable cell elongation, whereas oblique or longitudinal microtubules favour a shift in the growth axis towards lateral growth.
2. Control of cell division: the microtubular preprophase band defines axis and symmetry of the ensuing cell division. It marks the site where, after completion of chromosome segregation, the new cell plate will be laid down. This is the cellular basis for the control of branching patterns and phyllotaxis.
3. Control of cell-wall structure: cortical microtubules are bundled at those sites, where cell-wall thickenings are going to be formed. The orientation of cortical microtubules will therefore define the direction of these cell-wall thickenings and thus the spatial framework for lignification. This influences the mechanical properties of wood.

The second part of the book covers the role of microtubules in response to environmental factors. The focus is on three aspects of this vast field:

4. Control of the response to biotic stress: microtubules seem to be involved in the migration of the nucleus towards the infected site upon pathogen attack. The spread of plant viruses such as the tobacco mosaic virus between cells of infected plants seems to utilize actin microfilaments and microtubules. Reorientation of microtubules that are aligned over several cells accompanies wound healing and the establishment of new vessel contacts. Formation of mycorhiza and *Rhizobium*-induced root-nodules are further topics in this context.
5. Control of the response to metals: metal ions, such as aluminum or cadmium, limit crop yields in about 40% of the world's arable lands. They cause swelling of root cells and a loss of cell axis. For aluminum, a direct interaction with tubulin

dynamics has been discovered, opening the possibility to analyze and control the cytoskeletal response to this metal.

6. Control of the response to low temperature: microtubules depolymerize in response to chilling. In plants, the cold sensitivity of microtubules is well correlated with the chilling tolerance of the whole plant. It is possible to manipulate the cold sensitivity of microtubules by ecophysiological rationales and/or certain growth regulators. Moreover, various tubulin isotypes seem to exist that differ in cold sensitivity.

The third part of the book deals with the tools that can be used for biotechnological manipulation:

7. Tubulin genes: in all plant species tested so far there exist several tubulin genes corresponding to several tubulin isotypes with subtle differences in charge, tissue expression, temporal expression and signal inducibility. The corresponding tubulin isotypes seem to confer altered responses of microtubules to cold, herbicides and hormones. These isotypes could either be used directly to manipulate the behaviour of microtubules and thus the response of the plant to stress, or on the other hand, the promotors for these genes could be utilized to drive the expression of other genes of interest with a specific, possibly inducible, spatiotemporal pattern of expression.

8. Cytoskeletal mutants: an increasing panel of mutants becomes available that has been selected either for altered resistance to cytoskeletal drugs or for a changed pattern of morphogenesis.

9. Cytoskeletal drugs: several herbicides act either directly on microtubule assembly or indirectly on microtubule dynamics by interfering with signal chains that control microtubule dynamics. In addition, several growth regulators exert their effect via the microtubular cytoskeleton. There exist species and cultivar differences in drug sensitivity that could be used for weed control as well as for the control of crop growth.

The scope of the book is twofold: it gives a comprehensive overview of the numerous functions of microtubules during different aspects of plant life, and it proposes to make use of the potential of microtubules to influence fundamental aspects of plant life such different as height and shape control, mechanical properties of wood or resistance to pathogens or abiotic stress.

Freiburg, Germany, March 2000   Peter Nick

Contents

1 Control of Plant Height

Peter Nick
Institut für Biologie II, Schänzlestr. 1, D-79104 Freiburg, Germany

1.1
Summary

Control of plant height is important for agriculture, because plant height determines the mechanical stability of many crop plants. Resistance to wind, logging resistance, efficiency of light usage, microclimate, but also the partitioning of biomass from vegetative to reproductive structures are traits that depend directly or indirectly on plant height. In most cases, a reduction in plant height is the desired trait. However, conventional agriculture that is characterized by high density and high nutrient supply tends to create an environment enhancing plant height, and thus reducing plant stability (e.g. increased internode length caused by the shade avoidance response).

On the cellular level, these responses are based primarily on increased cell elongation. The repartitioning of cell expansion from longitudinal towards lateral expansion requires manipulation of the reinforcement mechanism that controls cell axiality. The orientation of cellulose microfibrils plays a pivotal role in this process. The deposition of cellulose and thus the axiality of the cell are controlled by cortical microtubules. The current state of the microtubule-microfibril syndrome will be briefly reviewed at this point (physical contact versus membrane channels). In the next step, the research on microtubule reorientation triggered by environmental signals will be presented with a special focus on those factors that are likely to be involved in the field (shade avoidance, phytochrome, gibberellins). The chapter will continue with the biochemical knowledge that has been accumulated so far about the mechanism of microtubule reorientation. In the final part, potential applications will be given: manipulation of microtubule dynamics by transgenic approaches, by application of certain chemicals, and by ecophysiological rationals.

1.2
Significance of the problem for agriculture

Control of plant height has been a major topic in agriculture for several decades, especially in graminean crops, such as wheat, rice, barley, rye and oats (Luib and Schott 1990; Borner et al. 1996). In wheat and barley, the resistance of a plant to lodging and windbreak has been found to be inversely related to plant height (Oda et al. 1966):

$$L_R = \frac{W \cdot M}{l^2 \cdot w}$$

with w=fresh weight, M=bending momentum at breaking, l=culm height and w=dry weight of the culm per tiller. Thus, lodging resistance will increase parabolically with decreasing plant height.

The agricultural losses caused by lodging are enormous: up to 10-50% in wheat (Laude and Pauli 1956; Weibel and Pendleton 1961), up to 60% in barley (Schott and Lang 1977; Knittel et al. 1983), and 20-40% in rice (Basak 1962; Kwon and Yim 1986; Nishiyama 1986). In addition, the quality of the grains in terms of malting quality, baking quality or carbohydrate content are negatively affected by lodging (Andersen 1979). Moreover, the costs for harvest and grain drying may increase by 50% (Luib and Schott 1990).

Lodging occurs mainly in two forms, as so-called root lodging, a bending of the culms just above the surface (Pinthus 1973), or as so-called stem lodging at higher internodes after heading. Approaches to improve lodging resistance in graminean crops have focussed either on the introduction of dwarfing genes by conventional breeding (Borner et al. 1996; Makela et al. 1996; Mcleod and Payne 1996), or, alternatively, on the application of growth regulators such as chlormequat chloride or ethephone (Schott and Lang 1977; Schreiner and Reed 1908; Tolbert 1960). Both approaches often act via the gibberellin pathway. The majority of the known dwarfing genes are affecting either gibberellin biosynthesis (Phinney and Spray 1990; Borner et al. 1996) or the responsiveness to gibberellins (Borner et al. 1996). The growth-retardant chlormequat interferes with gibberellin biosynthesis (Frost and West 1977) or utilization (Coolbaugh and Hamilton 1976). Ethephon, on the other hand, acts by stimulating the release of endogenous ethylene (Andersen 1979).

These massive shifts in the hormonal balance, caused either genetically or by application of growth retardants, can have undesirable side effects, such as precocious flowering along with incomplete grain filling (Makela et al. 1996), increased transpiration and reduced drought resistance (Kirkham 1983), reduced gametic viability (Dotlacil and Apltauerova 1978) or reduced root development (Luib and Schott 1990). In addition, only some crops are amenable to this strategy (Jung 1964; Clark and Fedak 1977): whereas wheat and rye respond very well, the success is less pronounced in species such as barley, maize or rice. Moreover, the success of growth retardants seems to be limited to root lodging, whereas chemical suppression of stem lodging bears increased risks for unfavourable side effects (Andersen 1979). Despite these serious drawbacks of chemical growth retardants, they have been used extensively – in 1989, for instance, 46% of wheat, and 18% of barley fields in Europe were treated with growth retardants (Luib and Schott 1990).

The success of this strategy is further limited by the specific environment created by modern agriculture characterized by high nutrient influx and high canopy density. These conditions stimulate internode elongation and thus increase the susceptibility of the crops to lodging and windbreak (Luib and Schott 1990). Most of our crop plants behave as shade avoiders, i.e. they are obligate sun plants (Smith 1981). In dense canopies they exhibit a pronounced shade-avoidance response: they perceive the presence of their neighbours by sensing changes in the ratio between red and far-red light by the photoreversible plant photoreceptor phytochrome (Smith 1981). They respond to this change in red/far-red ratio by enhanced elongation of stems and petiols. The adaptive value of the shade-avoidance response is supposed to be a protection against overgrowth by neighbouring plants. In a dense canopy, such as a wheat field, this response is definitely undesired in terms of yield and lodging resistance. Field trials with tobacco plants overexpressing phytochrome, and thus incapable of sensing the reflected light, properly demonstrate that the suppression of the shade-avoidance response allows for an increased allocation of assimilates and a concomitant increase in yield (Robson et al. 1996).

Reduced plant height is a desirable trait in other crops as well: the mechanization of seedling transplantation in paddy rice cultivation requires short sturdy seedlings (Schott et al. 1984). Reduced internode length is correlated to higher yield in rape seed, probably due to an increased penetration of light into the canopy (Luib and Schott 1990). In fruit trees, reduced plant height is a prerequisite for mechanical picking (Luib and Schott 1990), and in cotton cultures, compact growth is desired to ensure access of light and insecticides to the lower parts of the plant (Luib and Schott 1990). It should be mentioned, however, that in a few cases, the desired trait is an increased length of internodes, for instance in sugar cane, where in cooler growth seasons yield seems to be limited by reduced stem elongation (Tanimoto and Nickell 1967; Coleman 1958).

The significance of biotechnological approaches to control plant height can be summarized in the following statements:

1. Reduced stem elongation is a prerequisite for lodging control.
2. Reduced stem elongation is often correlated with improved partitioning of biomass into the reproductive structures.
3. High density of canopies and high nutrient influx, conditions that are characteristic for agricultural ecosystems, favour increased stem elongation.
4. Strategies based on growth retardants face, in some cases, serious drawbacks due to limited applicability (stem lodging), limited responsiveness (rice, oat), or deleterious side effects (reduced rooting, gamete viability).
5. The economic impact of height control is enormous.

These statements illustrate that the development of new approaches to control plant height is required and useful. In order to develop such approaches, it is necessary to consider plant growth at the cellular level.

1.3
Cellular mechanisms of plant growth

Growth can be brought about by an increase either in cell number (division growth) or individual cell volumes (expansion growth). Division growth is characteristic for organisms or developmental periods with determinate morphogenesis such as animals, plant embryos or leaf primordia after initiation (Steeves and Sussex 1989). During most of their life cycle plants exhibit indeterminate morphogenesis (Nick and Furuya 1992) and grow predominantly by cell expansion. In some organs, such as hypocotyls (Lockhart 1960) or coleoptiles (Furuya et al. 1969; Nick et al. 1994), the growth response could even be attributed entirely to cell expansion alone.

Plant cells expand by increasing the volume of the vacuole that accounts for more than 90% of total cell volume in most differentiated cells. The driving force for this increase in volume is a gradient of water potential with a more negative water potential in cytoplasm and vacuole as compared to the apoplast (Kutschera 1991). The expansion of the vacuole eventually would result in infinite swelling and a burst of the cell were it not limited by the rigid cell walls. The importance of the cell wall for the integrity of plant cells can be easily demonstrated if protoplasts are placed in a hypotonic medium (Fig. 1.1a).

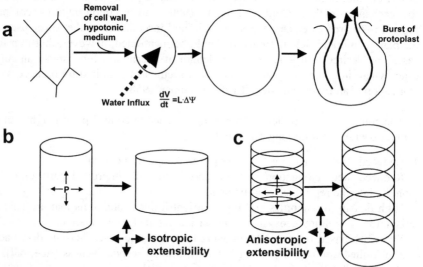

Fig. 1.1a-c.   Importance of the cell wall for limiting and guiding cell expansion. **a** Unlimited expansion of protoplasts in hypotonic medium. The volume increase per time is the product of the difference in water potential $\Delta\Psi$ between medium and protoplast and the conductivity L. **b** Lateral expansion in a cylindrical cell with cell walls of isotropic extensibility. **c** Elongation in a cylindrical cell, where lateral growth is inhibited by a reinforcement mechanism (transverse rings of cellulose).

Most plant cells derive directly or indirectly from isodiametric meristematic cells. Nevertheless, most differentiated cells in expanding tissues such as hypocotyls, internodes, petioles or coleoptiles are characterized by an approximately cylindrical shape. This cylindrical shape is usually lost upon removal of the cell wall – protoplasts are spherical with very few exceptions. These simple observations already demonstrate the importance of the cell wall for the control of cell shape.

In cylindrical cells, cell expansion is expected to occur preferentially in lateral direction, which should gradually corroborate the axiality of these cells (Fig. 1.1b). This means, on the other hand, that cylindrical cells must provide some kind of reinforcement mechanism to maintain their original axiality (Green 1980). This reinforcement mechanism seems to reside in the cell wall and was first described in the long internodal cells of the alga *Nitella* (Green and King 1966). In these elongate cells, the cellulose microfibrils were demonstrated by electron microscopy to be arranged in transverse rings, especially in the newly deposited inner layers of the wall. It should be mentioned that an anisotropic arrangement of cellulose had already been inferred from polarization microscopy much earlier, and the birefringency of the cell wall had been related to growth (Ziegenspeck 1948). However, the functional significance of this observation had not been recognized at that time. It is evident that the transverse arrangement of microfibrils can account for the reinforcement mechanism that maintains the longitudinal growth axis in cylindrical cells (Fig. 1.1c). This correlation between transverse microfibrils and cell elongation has been found in numerous examples and discussed in several reviews (Robinson and Quader 1982; Kristen 1985; Giddings and Staehelin 1991). Moreover, reorientations of the growth axis were found to be accompanied by a loss or a reorientation in the anisotropy of cellulose deposition (Green and Lang 1981; Lang et al. 1982; Hush et al. 1990). Therefore, the correlation between guided deposition of cellulose and cell growth seems to be very tight.

In intact organs, the control of growth axiality is not necessarily maintained actively in each individual cell, but is sometimes confined to specific tissues such as the epidermis, that are the targets for growth control. If coleoptiles or stem sections are split and subsequently allowed to grow in water, they usually curl inside out, i.e. the inner tissues grow faster than the epidermis. If growth promoting substances such as the heteroauxin indole-acetic acid are added, they begin to curl outside in, demonstrating that now the epidermis grows faster than the subtending inner tissue. This response has been found to be so sensitive that it could be used as a classical bioassay for auxin (Schlenker 1937). Biophysical measurements confirmed later that, in fact, auxin stimulates elongation of maize coleoptiles by increasing epidermal extensibility and thus releasing the constraint imposed on the elongation of the compressed inner tissues by the rigid epidermis (Kutschera et al. 1987). In this and many other cases it is the epidermis that is the target tissue for growth control but, generally, this example illustrates that it is important to manipulate the correct target cells, if one wants to control growth of intact plants efficiently.

1.4
The microtubule-microfibril interaction

As discussed above, the orientation of cellulose microfibrils is closely related to the preferential axis of cell expansion. In order to approach this phenomenon, it is necessary to consider briefly the synthesis of cellulose. New microfibrils are produced at cellulose-synthetizing enzyme complexes residing in the plasma membrane. These enzyme complexes, often designated terminal complexes, are usually organized in hexagonal configurations, the so-called rosettes (Giddings and Staehelin 1991). It is generally believed that these rosettes slide within the membrane, leaving behind them bundles of crystallizing cellulose, the microfibrils (Fig. 1.2). The mechanisms driving and guiding this sliding movement are expected to play a key role for the determination of the growth axis.

Cortical microtubules were predicted to exist (Green 1962) and to be responsible for guided cellulose deposition even before they were actually discovered electron microscopically by Ledbetter and Porter (1963). During subsequent years, an extensive body of information was accumulated, demonstrating an intimate link between cortical microtubules and the preferential axis of growth. The basic evidence is the following:

1. Cortical microtubules are closely associated with the plasma membrane, in plasmolyzing cells a direct contact between cortical microtubules and newly formed cellulose microfibrils has been detected by electron microscopy (Robinson and Quader 1982; Kristen 1985; Giddings and Staehelin 1991).
2. Preceding the formation of secondary wall thickenings the prospective sites of their formation are already marked by parallel thick bundles of cortical microtubules (Fukuda and Kobayashi 1989; Jung and Wernicke 1990).
3. In those cases where the preferential axis of cellulose deposition (and concomitantly the axis of cell growth) changes, this reorientation is heralded by a reorientation of cortical microtubules (ethylene response, Lang et al. 1982; auxin response, Bergfeld et al. 1988; gibberellin response, Toyomasu et al. 1994; wood formation, Abe et al. 1995).
4. Elimination of cortical microtubules by antimicrotubular drugs results in a gradual loss of growth anisotropy and a block of cell elongation leading, in extreme cases, to lateral swelling (Hogetsu and Shibaoka 1978; Robinson and Quader 1981; Kataoka 1982; Bergfeld et al. 1988; Vaughan and Vaughn 1988; Nick et al. 1994; Baskin and Bivens 1995).

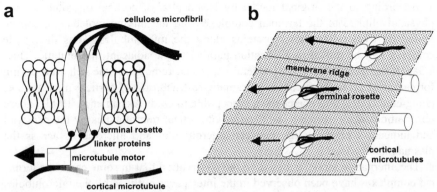

Fig. 1.2. Organization of cellulose-synthetizing enzymes in the plasma membrane. The complexes are arranged in rosette-like terminal complexes and the cellulose microfibrils are secreted towards the apoplastic side of the membrane. By directional movement of the terminal complexes within the membrane a directional deposition of cellulose is achieved.

The exact mechanism by which microtubules drive and guide cellulose deposition has been under debate since the discovery of cortical microtubules by Ledbetter and Porter (1963), and manifold different hypotheses have been proposed (Robinson and Quader 1982; Kristen 1985; Giddings and Staehelin 1991). The principal debate can be summarized into two alternative models:

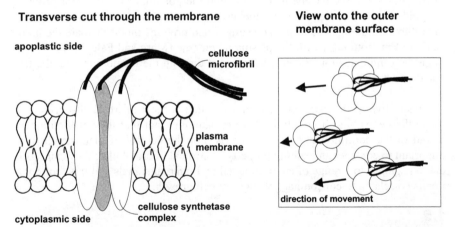

Fig. 1.3a,b. Cortical microtubules guide cellulose deposition. **a** According to the model by Heath (1974), a microtubule-based motor protein is linked physically to the terminal complexes. The driving force for the movement is provided by this protein, the direction of the movement is determined by the direction of cortical microtubules. **b** According to the model by Giddings and Staehelin (1991), the driving force for the movement is derived from the crystallization of cellulose, the direction of the movement is determined by membrane channels that form between cortical microtubules.

1. According to the original model by Heath (1974), cortical microtubules are physically linked to the terminal complexes and the linking molecule(s) can be pulled by dynein-like motor proteins along the microtubules. Thus the whole complex will be pulled in a direction parallel to the adjacent microtubules (Fig. 1.3a). It has been observed in several cases that removal of the cell wall during formation of protoplasts causes a dramatic restructuring of cortical microtubules (Jung et al. 1993) and makes them susceptible to cold (Akashi et al. 1990). These observations demonstrate a stabilization of cortical microtubules by the cell wall and indicate a physical link between microtubules and microfibrils across the plasma membrane.

2. The alternative model is based on the observation that in some cases the terminal complexes have been observed in the interspaces outlined by the microtubules rather than being directly attached to them (Giddings and Staehelin 1991). The guiding of rosette movement, according to this model, is not caused by a physical link of the terminal complexes to microtubule motors. Microtubules are rather supposed to induce membrane channels that impede lateral deviations of rosette movement (Fig. 1.3b). The driving force for the movement would be cellulose crystallization itself, propelling the terminal rosette through the microtubule-dependent membrane channels.

At the present stage, it is difficult to decide between these two models. Moreover, neither of them seems to be complete and able to accommodate all observations. It is necessary to understand, on the molecular level, the interaction between microtubules and the plasma membrane, and the potential role of motor proteins for guided cellulose deposition. In this context, the recent discovery of a 90-kDa microtubule-binding protein is interesting. Such proteins might mediate the association of microtubules with the plasma membrane (Marc and Palevitz 1990) and might be candidates for proteins that interact through the membrane with the terminal complexes.

Irrespective of the exact mechanism by which cortical microtubules drive and guide the deposition of cellulose, there remains the fact that they are intimately linked to the definition of the growth axis. For the control of plant height it is important to consider an exciting property of cortical microtubules: they can reorient in response to a range of environmental and internal signals and this reorientation is correlated to concomitant changes of cell growth.

1.5
Signal-triggered reorientation of microtubules

When seedlings of higher plants encounter mechanical obstacles, they display a characteristic barrier response involving a shift of the growth axis from elongation towards stem thickening. The trigger for this response seems to be the plant hormone ethylene (Nee et al. 1978), that is constantly formed by the growing stem and accumulates in front of physical obstacles. This ethylene-induced block of internode elongation with a subsequent thickening of the stem is the mechanism by which ethephone can increase lodging resistance (Andersen 1979).

Using the ethylene-triggered switch of the growth axis, Lang et al. (1982) succeeded in demonstrating that environmental signals can control growth via the microtubule-microfibril pathway. These authors observed, by electron microscopy in pea epicotyls, that cortical microtubules reorient from their original transverse array into steeply oblique or even longitudinal arrays. This reorientation is followed by a shift of cellulose deposition from transverse to longitudinal and subsequently a thickening of the stem.

In subsequent years, similar correlations between growth, microfibril deposition, and cortical microtubules could be shown for other hormones as well. In coleoptile segments of maize, elongation is under control of auxin and limited by the epidermal extensibility (Kutschera et al. 1987). Target for the action of auxin is the outer epidermal cell wall. Auxin increases the extensibility of this wall (so-called wall loosening) and thus releases the restraint put upon the extension of the inner tissues by the epidermis. In segments that have been depleted of endogenous auxin, cortical microtubules and newly deposited cellulose microfibrils have been observed by electron microscopy to be oriented in longitudinal direction (Bergfeld et al. 1988). They both are found to be transverse after addition of exogenous indoleacetic acid in parallel with a restoration of elongation growth.

With the adaptation of immunofluorescence microscopy to plant cells (Lloyd et al. 1980), it became possible to investigate not only the factors that can trigger reorientation of cortical microtubules, but also to follow the dynamics of this process. These studies revealed that not only various plant hormones such as auxins (Bergfeld et al. 1988; Nick et al. 1990, 1992; Nick and Schäfer 1994), gibberellins (Mita and Katsumi 1986; Nick and Furuya 1993; Sakiyama-Sogo and Shibaoka 1993; Shibaoka 1993; Toyomasu et al. 1994) and abscisic acid (Sakiyama-Sogo and Shibaoka 1993) can induce reorientation of microtubules, but also physical factors such as blue light (Nick et al. 1990; Laskowski 1990; Zandomeni and Schopfer 1993), red light (Nick et al. 1990; Nick and Furuya 1993; Zandomeni and Schopfer 1993; Toyomasu et al. 1994), gravity (Nick et al. 1990, 1997; Blancaflor and Hasenstein 1993; Nick et al. 1997), high pressure (Cleary and Hardham 1993), mechanical stress (Zandomeni and Schopfer 1994), wounding (Hush et al. 1990) and electrical fields (Hush and Overall 1991).

However, only in a few cases has the dynamics of signal-triggered reorientation of microtubules been investigated in the context of signal-induced changes of growth. In maize coleoptiles, microtubules were observed to reorient rapidly from transverse to longitudinal upon phototropic stimulation (Nick et al. 1990). This reorientation was confined to the lighted flank of the coleoptile and preceded phototropic curvature. The time course for the auxin-dependent reorientation in the same system supported a model (Fig. 1.4), where photo- or gravitropic stimulation induced a displacement of auxin transport from the lighted flank of the coleoptile to the shaded flank. The depletion of auxin in the lighted flank subsequently induces a reorientation of cortical microtubules from transverse to longitudinal (Nick et al. 1990), and in parallel a longitudinal deposition of cellulose microfibrils (Bergfeld et al. 1988). In the auxin-enriched shaded flank both microtubules and microfibrils remain transverse. The gradient of microfibril orientation should then result in a decrease in longitudinal extensibility of epidermal cell walls in the lighted side and, eventually, in asymmetric growth leading to phototropic curvature towards the light pulse.

A detailed investigation of this phenomenon revealed, however, a more complex reality (Nick et al. 1991, 1992; Nick and Schäfer 1994; Nick and Furuya 1996). It is possible, by rotating the seedlings on a clinostat, and without preceding tropistic stimulation to obtain a so-called nastic bending that is not accompanied by a gradient in the orientation of cortical microtubules across the coleoptile cross-section (Nick et al. 1991). On the other hand, the gradient of microtubule orientation induced by a phototropic pulse stimulation persists, whereas the phototropic curvature vanishes due to gravitropic straightening (Nick et al. 1991). In parallel to the phototropic curvature, a phototropic stimulus induces a longlasting transverse polarization of the tissue that is persistent over several days. This polarity can cause stable changes in growth (Nick and Schäfer 1988, 1991, Nick and Schäfer 1994) and can even control morphogenetic events such as the emergence of crown-roots that occurs several days after the stimulus had been administered (Nick 1997). These stable changes in growth are closely related to an irreversible fixation of microtubule orientation (Nick and Schäfer 1994).

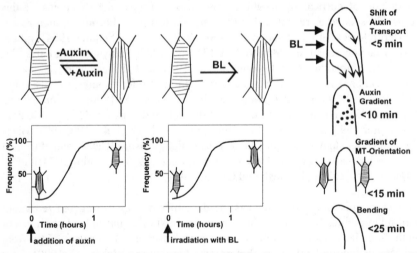

Fig. 1.4. Behaviour of cortical microtubules during phototropic curvature of maize coleoptiles. (Nick et al. 1990). Microtubules reorient from transverse to longitudinal in response to auxin depletion or in response to phototropic stimulation. The reorientation induced by phototropic stimulation is confined to the lighted flank of the coleoptile and initiates subsequent to the auxin displacement across the coleoptile, but prior to the onset of phototropic curvature.

Two hours after phototropic stimulation, cortical microtubules lose their ability to reorient in response to a light pulse, at the same time as the transverse polarity detectable on the physiological level becomes persistent. Interestingly, microtubules lose their ability to respond to auxin as well, indicating that the fixation of their orientation is not caused by sensory adaptation of phototropic perception (Nick and Schäfer 1994). This fixation of microtubules requires blue light, and the effect of light cannot be mimicked by depletion of auxin or by gradients of auxin depletion.

These studies suggest that the microtubule-microfibril pathway is responsible for persistent long-lasting changes in growth. There must be, in addition, a second pathway that can control growth independently, and that seems to be responsible for fast growth responses. However, this second unknown pathway has been detected only in careful kinetic studies, if discrepancies between growth and microtubule reorientation are explicitly investigated. In this context it should be mentioned that in some cases the microtubule response has been found to be somewhat slower than the signal-triggered response of growth: blue-light-induced growth inhibition of pea stems (Laskowski 1990), gravitropism of maize roots (Blancaflor and Hasenstein 1993). Considering the mechanism of growth control, it is evident that the microtubule-microfibril pathway is designed for persistent changes in growth: it requires a certain time until enough cellulose microfibrils are deposited in the new direction (Lang et al. 1982) before a change in growth can occur. Rapid growth responses require a control mechanism that is expected to act independently of cellulose deposition. The irreversible fixation of microtubule orientation (Nick and Schäfer 1994) in response to blue light indicates that both pathways of growth control act in parallel and play complementary roles. If persistent changes in growth are to be achieved, the microtubule-microfibril pathway is a good target for biotechnological application. To design approaches for artificial manipulation of this pathway, it is necessary to consider the potential mechanism of signal-triggered microtubule reorientation.

1.6
How do microtubules reorient?

Immunofluorescence microscopy allowed investigation of the plant cytoskeleton in its three-dimensional organization for the first time (Lloyd et al. 1980). These studies revealed that, in elongating cells, the microtubules are arranged in helicoidal arrays along the periphery of the cell. This observation stimulated the first model for microtubule reorientation (Lloyd and Seagull 1985): the helicoidal arrays were perceived as dynamic springs with variable pitch. If microtubules are forming this helix slide in such a way that the helix is shortened, this will result in a steep pitch and in longitudinal microtubules (Fig. 1.5a). If they slide in the opposite direction, the spring will relax, resulting in an almost transverse pitch. According to this model, the molecular mechanism of reorientation is expected to involve microtubule motors, such as kinesin or dynein.

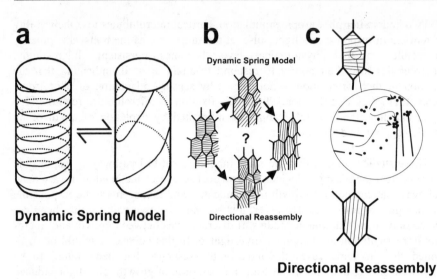

a

Dynamic Spring Model

b

Dynamic Spring Model

?

Directional Reassembly

c

Directional Reassembly

Fig. 1.5a-c. Potential mechanisms for the reorientation of cortical microtubules. **a** According to the dynamic-spring model by Lloyd and Seagull (1985), cortical microtubules are organized into a mechanically coupled helix. By mutual sliding of microtubules the helix can either relax resulting in an almost transverse pitch, or it can contract resulting in an almost longitudinal pitch. **b** Transitions between transverse and longitudinal microtubules – the dynamic-spring model would predict homogeneously oblique microtubules as transitional state; in fact, a patchwork of transverse microtubules coexisting with longitudinal microtubules is observed consistent with the directional-reassembly model. **c** According to the directional-reassembly model, the equilibrium between assembly and disassembly is direction-dependent, resulting in a net reorientation of microtubules and a transitional state, where transverse and longitudinal microtubules coexist.

This model is, in fact, very attractive in its simplicity and elegance. However, an increasing body of information questions the validity of the dynamic-spring model. This evidence can be summarized as follows:

1. According to the dynamic-spring model, the cortical microtubules are mechanically coupled and comprise a physical entity. In epidermal tissues, the reorientation of cortical microtubules is confined to the microtubules adjacent to the outer wall (Bergfeld et al. 1988; Nick et al. 1990; Wymer and Lloyd 1996), leading to a situation with transverse microtubules at the inner wall and longitudinal microtubules at the outer wall. This difference in orientation is difficult to reconcile with the concept of a mechanically coupled spring.

2. The transitions between transverse and longitudinal arrays of microtubules should involve situations where microtubules are homogeneously oblique and then gradually change pitch until the longitudinal array is reached. Although oblique microtubules can be observed, they seem to occur rather as a final than a transitional state (Gunning and Hardham 1982; Hush et al. 1990). Early phases of reorientation, triggered by strong stimuli or incomplete reorientation in response to suboptimal stimulation, tend to produce a different type of transition (Nick et al. 1990, 1992). In contrast to the situation predicted by the dynamic-spring

model, a patchwork of transverse and longitudinal microtubules is observed, with transverse and longitudinal microtubules being interspersed even within the same cell (Fig. 1.5b).

3. If microtubule depolymerization is suppressed by addition of taxol, an inhibition of microtubule reorientation has been observed in several cases (Falconer and Seagull 1985; Nick et al. 1997), indicating that microtubule depolymerization is required for reorientation. This is not expected from the dynamic-spring model. It should be mentioned, however, that this inhibition of taxol was not found in protoplasts (Wymer et al. 1996).

4. Cortical microtubules were originally thought to be relatively inert lattices. Recent experiments involving microinjection of fluorescent-labelled neurotubulin into living epidermal cells (Yuan et al. 1994; Wymer and Lloyd 1996) revealed that the injected tubulin was incorporated extremely rapidly into the preexisting cortical network, and that upon bleaching of fluorescence by a laser beam, the fluorescence of the bleached spot recovered within a few minutes, indicating an extremely high turnover of tubulin monomers. This high dynamics of tubulin assembly and disassembly contrasts with the concept of a mechanically coupled cytoskeletal helix.

5. The observation of microtubule reorientation in vivo in epidermal cells that were microinjected with fluorescent-labelled neurotubulin (Yuan et al. 1994; Wymer and Lloyd 1996) demonstrates local phase transitions from transverse to longitudinal arrays as first steps of reorientation, resulting in stochastically arranged patches of transverse and longitudinal microtubules. These patches subsequently extend, eventually merging into a homogeneously longitudinal array.

These observations suggest an alternative mechanism of microtubule reorientation: direction-dependent assembly and disassembly (Fig. 1.5c). Dependent on the orientation of a given microtubule, it might be in a growing state (assembly dominating over disassembly) or in a shrinking state (so-called catastrophe with disassembly dominating over assembly). If this model is correct, the key for manipulating microtubule reorientation has to be sought among those factors that control assembly and disassembly of microtubules.

The assembly of microtubules is promoted in vitro by warm temperature, by GTP and by magnesium. In vivo the nucleation of new microtubules occurs on the surface of specialized organelles, the centrosomes. Higher plants are unique in this respect, because they do not possess centrosomes. They do, however, possess however, functional analogues, so-called microtubule-organizing centers (MTOCs). In dividing cells a major MTOC seems to be the nuclear surface (Lambert 1993) and has been shown to induce the formation of new microtubules (Stoppin et al. 1994). In epidermal cells, cortical MTOCs have been described where new microtubules are formed during the recovery from depolymerization induced by drugs, low temperature or high pressure (Marc and Palevitz 1990; Cleary and Hardham 1993). The molecular composition of these MTOCs is not understood, but it seems that epitopes that are present in centrosomes such as γ-tubulin (Liu et al. 1994) can be detected in plant MTOCs.

The answer to the question of where new microtubules are nucleated would explain the problem of reorientation only partially, however. It remains to be elucidated how the elongation of a microtubule is controlled in space and how the microtubule is interconnected to other microtubules, to other cytoskeletal elements and to the plasma membrane. These links are practically unknown. Two plant microtubule-associated proteins have been cloned so far, both of them being factors required for protein translation. One of these factors, EF-1α (Durso and Cyr 1994) has been described as bundling of microtubules in vitro, although it is not clear, whether it has this function in vivo. The other microtubule-associated protein, IF-(iso)4F, has been shown to induce end-to-end annealing of microtubules in vitro (Hugdahl et al. 1995). Again, the function of this protein in vivo is still unknown. Recently, a 90-kDa protein has been discovered (although not identified) that might be important for the interaction between microtubules and the plasma membrane (Marc et al. 1996).

The identification and cloning of these proteins are necessary steps for molecular approaches to microtubule reorientation. It should be kept in mind, however, that other mechanisms may contribute as well to direction-dependent microtubule stability, such as posttranslational modifications (Sullivan 1988). For instance, the pattern of α-tubulin-tyrosination changes during gibberellin-induced microtubule reorientation in dwarf pea (Duckett and Lloyd 1994). It will be important to resolve whether such posttranslational modifications are cause or consequence of altered microtubule stability – it is conceivable that the increased lifetime of stable microtubules allows for prolonged activity of modifying enzymes, resulting in a higher degree of modification.

In addition to posttranslational modifications, changes in the pattern of microtubule-associated proteins can be observed in response to light treatments that induce a reorientation of microtubules and, concomitantly changes in growth rate. In maize coleoptiles, a protein of 50 kDa apparent molecular weight is induced during phytochrome-dependent cell elongation and becomes colocalized with bundled arrays of cortical microtubules (Nick et al. 1995). This protein can bind to preformed microtubules, but it can also coassemble with tubulin dimers into microtubules and is found to be associated with the plasma membrane.

The molecular identification of these MAPs and modifying enzymes is interesting at its own value, but it does not touch the question of direction. Why are the microtubules disassembled in one direction, but assembled in the other? This direction-dependent stability is a factor that cannot be intrinsic to microtubules themselves. There must be some kind of either lattice or field that is responsible for the directional component of microtubule dynamics. This lattice might be either a different component of the cytoskeleton (e.g. actin microfilaments), physical fields (e.g. mechanical strains, bound dipoles) or apoplastic lattices (e.g. cell-wall components).

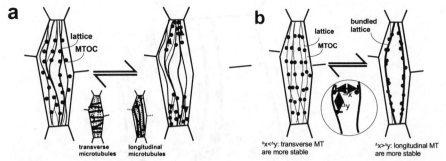

Fig 1.6a,b. Potential mechanisms responsible for directional reassembly of microtubules. **a** Microtubule-organizing centers could move along a directional matrix/lattice resulting in a redistribution of microtubule-nucleating sites. **b** Microtubule-stabilizing factors are organized along a directional matrix/lattice and this lattice is reorganized without a movement of these microtubule-stabilizing factors along the lattice. If the lattice becomes bundled, longitudinal microtubules will be favoured over transverse microtubules due to a lower minimal distance between the microtubule-stabilizing factors.

This lattice or field somehow participates in the spatial organization of microtubule nucleation and, possibly, elongation. One might think about MTOCs being moved and organized along such a lattice (Fig. 1.6a). It is more difficult to conceive a model explaining the control microtubule elongation. One possibility could be a spatial arrangement of microtubular bundling or cross-linking proteins that supports microtubule stability in a certain direction but not in the other, or a lattice that favours a certain orientation by creating a more favourable minimal distance between such supporting molecules (Fig. 1.6b). It is conceivable that this lattice consists of a stable subpopulation of microtubules themselves that would not be seen upon microinjection of fluorescent-labelled tubulin into living cells (Yuan et al. 1994). Although the molecular basis for microtubule reorientation has to be elucidated, it is already possible at this stage to develop general designs for microtubule-based approaches towards a control of plant height.

1.7
Potential approaches for manipulation

The control of length growth by environmental stimuli such as light, gravity or nutrients involves the following steps (Fig. 1.7): (1) stimulus perception, (2) signal transduction cascades that are often organized as complex networks, (3) changes in an unknown directional matrix that defines direction-dependent microtubule disassembly/assembly, (4) stimulation of microtubule disassembly, (5) stimulation of microtubule assembly (dependent on the direction of microtubules, resulting in a net reorientation of cortical microtubules and (6) reorientation of cellulose deposition, leading to a directional switch in the anisotropy of the cell wall, and a switch in the axis of preferential cell expansion.

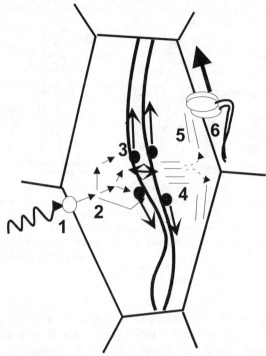

Fig. 1.7   Potential cellular targets for length control. **1** Signal perception; **2** signal transduction cascades and networks; **3** movement of microtubule-organizing centers and/or reorganization of the directional matrix linked to microtubule-nucleating and/or microtubule-elongating factors; **4** disassembly of transverse microtubules; **5** assembly of longitudinal microtubules; **6** microtubule-guided movement of cellulose-synthetizing enzyme complexes.

In the following, a potential manipulation of these steps will be discussed individually for each of these steps.

1. Manipulation of stimulus perception. The dense canopies characteristic or modern agriculture lead to a pronounced shade avoidance response, triggered by changes in the activity of the plant photoreceptor phytochrome (Smith 1981). The stimulation of stem elongation caused by shade avoidance contributes to a increased probability for lodging and thus to a reduction in yield (Oda et al. 1966). It is possible to block the shade-avoidance response by overexpression of the photoreceptor (phytochrome), and thus to increase yield in some cases (Robson et al. 1996). It has to be kept in mind, however, that the consequences of this strategy are not confined to elongation, but are expected to affect all responses that are triggered by phytochrome, such as the composition of photosystems, branching, tropistic responses, hormonal balance and induction of flowering. Depending on species and type of crop, the overexpression of phytochrome might cause undesired side effects that are difficult to foretell. The same type of argumentation is valid not only for shade avoidance, but also for manipulation of other pathways

triggered, for instance, by temperature, nutrient uptake or abiotic and biotic stresses.

2. Manipulation of signal-transduction cascades. The classical approaches for length control using dwarfing genes or growth regulators such as CCC and ethephone principally fall into this class (Andersen 1979). Most signal chains are split and interwoven into complex networks. Our knowledge about these interactions is still rudimentary for most plant systems and the available tools are still relatively crude, usually allowing only for global shifts of the network, for instance by changing the hormonal balance. This relatively "rude" approach leads to a high degree of pleiotropy and is likely to cause undesired side effects, as for instance, incomplete grain filling (Makela et al. 1996) or reduced root development (Luib and Schott 1990). As long as the complex mutual interactions between signal chains are poorly understood, the strategy must be to search the target for manipulation downstream of the branching points where the signal cascades split into parallel chains.

3. Manipulation of the directional matrix. This target might be the most specific, because it is expected to affect cortical microtubules in the first place and thus mainly the axis of cell expansion. Unfortunately, the nature of this directional matrix is still completely unknown. Experiments involving the microfilament blocker cytochalasin D indicate, however, that this matrix might be related to the actin cytoskeleton (Kobayashi et al. 1988; Seagull 1990; Nick et al. 1997).

4. Manipulation of microtubule disassembly and reassembly. This step is, so far, is certainly the most accessible to manipulation. Disassembly of microtubules can be inhibited by various drugs such as taxol (Parness and Horwitz 1981) and it can be stimulated by a range of compounds such as the alkaloids colchicin, vinblastine and colcemid or herbicides of the dinitroaniline group (Morejohn 1991). Microtubule disassembly can be triggered also by cold and the calcium-calmodulin pathway. Stabilization of microtubules can be achieved by certain microtubule-binding proteins such as EF-1α (Durso and Cyr 1994), components of the cell wall (Akashi et al. 1990; Marc et al. 1996), certain hormones such as abscisic acid (Sakiyama and Shibaoka 1990) or by environmental stimuli such as blue light (Nick and Schäfer 1994). There exist several possibilities how these tools could be used for a strategy to control length growth. There are two key aspects that will be discussed below: specificity of the approach and possibilities for amplitude modulation.

6. Manipulation of cellulose synthesis. It is not clear whether and how microtubules are linked to the terminal complexes responsible for cellulose synthesis. If microtubules are mechanically coupled to the cellulose-synthase rosettes (Heath 1974), one might try to block or reestablish this link. This would be a very elegant way to uncouple cell expansion from the environment or, if necessary, to restore signal control of growth. If, however, the guidance of cellulose deposition by microtubules is brought about by microtubule-induced membrane channels (Giddings and Staehelin 1991), it is expected to be independent of physical links between microtubules and terminal complexes. In this case, the direction of cellulose deposition could be manipulated only indirectly by manipulation of microtu-

bule direction (see above). An inhibition of growth as such by blocking cellulose synthesis by appropriate inhibitors (Edelmann et al. 1989) would result in a thinning of cell walls and reduced mechanical stability. It might be feasible, however, to increase wall stiffness by induction of peroxidase-triggered lignification of the cell wall (Fry 1979).

In the following, a manipulation of microtubule assembly and disassembly will be considered in more detail with focus on the problems of specificity and amplitude modulation.

A. Specificity. The manipulation has to be specific in time and space. An overall manipulation of tubulin dynamics is expected to cause severe side effects such as changes in cell division patterns and frequency, sterility or even lethality. In the ideal case, the manipulation should be targeted to the tissue, where length growth is controlled – in most cases the epidermis (see above). If the approach is based on the application of certain drugs, a certain targeting to the epidermis is already achieved by the limited penetration of the drugs into the tissue. For transgenic strategies two principal approaches exist: first, the level of tubulin subunits could be downregulated in specific organs or even tissues using the respective antisense constructs based on internode or stem-specific tubulin promotors (see Breviario, Chapter 8). The reduction of microtubule number in the target cells should result in a gradual loss of cell-wall anisotropy and a repartitioning of growth from elongation towards stem thickening. Second, the dynamics of assembly and disassembly could be elevated or reduced in specific organs or tissues. For this, genes coding for microtubule-associated proteins that confer microtubule bundling (Durso and Cyr 1994) or membrane association (Nick et al. 1995; Marc et al. 1996) could be cloned and expressed under the control of internode-specific promotors, resulting in a fixation of microtubule orientation in the target tissue. A further level of specificity could be introduced by alteration of those gene products that mediate the interaction of certain signal pathways with tubulin disassembly. These genes are so far not known, but the existence of rice mutants where the microtubule reorientation in response to auxin is disturbed, whereas it can proceed in response to other stimuli (Nick et al. 1994), suggests that such gene products will become accessible in the future.

B. Amplitude manipulation. In order to design methods that are of practical use, a way must be introduced to modulate the strength of growth control. It is not feasible, for instance, to suppress cell elongation completely. A modulation could be achieved either by changing the strength of the driving promotor in transgenic rationals or by the introduction of a temporal component. The activity of the transgenic construct could be engineered under the control of a simple inductor that would allow growth to be manipulated at specific time points. In this context, naturally occurring inductors such as temperature or light should be considered as well, especially if they are changing rhythmically – a rhythmic inductor could control growth in each newly formed internode during a specific period of development (permitting a manipulation of cell elongation independently of cell division). In addition to chemical and transgenic rationales, the possibility of rela-

trigger microtubule depolymerization by low temperatures (Akashi et al. 1990) and to induce a fixation of microtubule orientation in longitudinal direction by irradiation with blue light (Nick and Schäfer 1994). In some crops that are precultivated in greenhouses before field planting (e.g. rice, tomatoes, many ornamental plants), it might be feasible to design special protocols of lighting and temperature that cause the desired effect on length growth. It should be mentioned in this context that such approaches have been already successfully used in the case of bulb forcing of ornamental plants such as tulips or hyacinths (Gude and Dijkema 1993).

The next years will extend and deepen our knowledge about those factors that are responsible for the assembly and disassembly of microtubules. The molecular characterization of these factors will open the possibility of controlling the behaviour of microtubules in specific tissues, at specific times, and in response to specific stimuli. The more specific these approaches, the more efficient they will be. Length control is a central topic in agriculture, and therefore microtubule-based rationales designed to regulate the axis of growth are expected to be a meaningful biotechnological strategy.

References

Abe H, Funada R, Imaizumi H, Ohtani J, Fukazawa K (1995) Dynamic changes in the arrangement of cortical microtubules in conifer tracheids during differentiation. Planta 197: 418-421

Akashi T, Kawasaki S, Shibaoka H (1990) Stabilization of cortical microtubules by the cell wall in cultured tobacco cells. Effects of extensin on the cold stability of cortical microtubules. Planta 182: 363-369

Andersen AS (1979) Plant growth retardants: present and future use in food production. NATO ASI Ser USA 22: 251-277

Basak MN (1962) Nutrient uptake by the rice plant and its effect on yield. Agron J 54: 373-376

Baskin TI, Bivens NJ (1995) Stimulation of radial expansion in *Arabidopsis* roots by inhibitors of actomyosin and vesicle secretion but not by various inhibitors of metabolism. Planta 197: 514-521

Bergfeld R, Speth V, Schopfer P (1988) Reorientation of microfibrils and microtubules at the outer epidermal wall of maize coleoptiles during auxin-mediated growth. Bot Acta 101: 57-67

Blancaflor EB, Hasenstein KH (1993) Organization of cortical microtubules in graviresponding maize roots. Planta 191: 230-237

Borner A, Plaschke J, Korzun V, Worland AJ (1996) The relationship between the dwarfing genes of wheat and rye. Euphytica 89: 69-75

Clark RV, Fedak G (1977) Effects of chloromequat on plant height, disease development and chemical constituents of cultivars of barley, oats and wheat. Can J Plant Sci 57: 31-36

Cleary AL, Hardham AR (1993) Pressure-induced reorientation of cortical microtubules in epidermal cells of *Lolium rigidum* leaves. Plant Cell Physiol 34: 1003-1008

Coleman RE (1958) The effect of gibberellic acid on growth of sugarcane. Sugar J 20: 23

Coolbaugh RC, Hamilton R (1976) Inhibition of *ent*-kaurene oxidation and growth by α-cyclopropyl α(p-methoxyphenyl-)-5-pyrimidine methyl alcohol. Plant Physiol 57: 245-248

Dotlacil L, Apltauerova M (1978) Pollen sterility induced by ethrel and its utilization in hybridization of wheat. Euphytica 27: 353-360

Duckett CM, Lloyd CW (1994) Gibberellic acid-induced microtubule reorientation in dwarf peas is accompanied by rapid modification of an a-tubulin isotype. Plant J 5: 363-372

Durso NA, Cyr RJ (1994) A calmodulin-sensitive interaction between microtubules and a higher plant homolog of elongation factor 1α. Plant Cell 6: 893-905

Edelmann H, Bergfeld R, Schopfer P (1989) Role of cell-wall biogenesis in the initiation of auxin mediated growth in coleoptiles of *Zea mays* L. Planta 179: 486-494

Falconer MM, Seagull RW (1985) Xylogenesis in tissue culture: taxol effects on microtubule reorientation and lateral association in differentiating cells. Protoplasma 128: 157-166

Frost RG, West CA (1977) Properties of kaurene synthesis from *Marah macrocarpus*. Plant Physiol 59: 22-29

Fry SC (1979) Phenolic components of the primary cell wall and their possible role in the hormonal regulation of growth in spinach. Planta 146: 343-351

Fukuda H, Kobayashi H (1989) Dynamic organization of the cytoskeleton during tracheary-element differentiation. Dev Growth Differ 31: 9-16

Furuya M, Pjon CJ, Fujii T, Ito M (1969) Phytochrome action in *Oryza sativa* L. III. The separation of photoperceptive site and growing zone in coleoptiles, and auxin transport as effector syste. Dev Growth Differ 11: 62-75

Giddings TH, Staehelin A (1991) Microtubule-mediated control of microfibril deposition. A re-examination of the hypothesis. In: Lloyd CW (ed) The cytoskeletal basis of plant growth and form. Academic Press, London, pp 85-99

Green PB (1962) Mechanism for plant cellular morphogenesis. Science 138: 1404-1405

Green PB (1980) Organogenesis – a biophysical view. Annu Rev Plant Physiol 31: 51-82

Green PB, King A (1966) A mechanism for the origin of specifically oriented textures with special reference to *Nitella* wall texture. Aust J Biol Sci 19: 421-437

Green PB, Lang JM (1981) Towards a biophysical theory of organogenesis: birefringence observations on regenerating leaves in the succulent *Graptopetalum paraguayense*. Planta 151: 413-426

Gude H, Dijkema M (1993) Light quality as a tool to improve bulbous flower quality. Eur Symp Photomorphogenesis in Plants, Tirrenia, July 11-15, 1993

Gunning BES, Hardham AR (1982) Microtubules. Annu Rev Plant Physiol 33: 651-698

Heath IB (1974) A unified hypothesis for the role of membrane-bound enzyme complexes and microtubules in plant cell wall synthesis. J Theor Biol 48: 445-449

Hogetsu T, Shibaoka H (1978) Effects of colchicine on cell shape and on microfibril arrangement in the cell wall of *Closterium acerosum*. Planta 140: 15-18

Hugdahl JD, Bokros CL, Morejohn LC (1995) End-to-end annealing of plant microtubules by the p86 subunit of eukaryotic initiation factor-(iso)4F. Plant Cell 7: 2129-2138

Hush JM, Overall RL (1991) Electrical and mechanical fields orient cortical microtubules in higher plant tissues. Cell Biol Int Rev 15. 551-560

Hush JM, Hawes CR, Overall RL (1990) Interphase microtubule re-orientation predicts a new cell polarity in wounded pea roots. J Cell Sci 96: 47-61

Jung G, Wernicke W (1990) Cell shaping and microtubules in developing mesophyll of wheat (*Triticum aestivum* L.). Protoplasma 153: 141-148

Jung G, Hellmann A, Wernicke W (1993) Changes in the density of microtubular networks in mesophyll cells derived protoplasts of *Nicotiana* and *Triticum* during leaf development. Planta 190: 10-16

Jung J (1964) Über die halmverkürzende Wirkung von Chlorcholinchlorid (CCC) bei Weizen und deren Abhängigkeit von der Bodenart. Z Pflanzenernähr Düng Bodenkd 107: 147-153

Kataoka H (1982) Colchicine-induced expansion of *Vaucheria* cell apex. Alteration from isotropic to transversally anisotropic growth. Bot Mag Tokyo 95. 317-330

Kirkham MB (1983) Effect of ethephon on the water status of a drought-resistant and a drought sensitive cultivar of winter wheat. Z Pflanzenphysiol 112: 102-112

Knittel H, Lang H, Schott PE, Höppner P (1983) Verbesserung der Standfestigkeit – auch bei Gerste und Roggen. BASF Mitt Landbau 2: 1-31

Kobayashi H, Fukuda H, Shibaoka H (1988) Interrelation between the spatial disposition of actin filaments and microtubules during the differentiation of tracheary elements in cultured *Zinnia* cells. Protoplasma 143: 29-37

Kristen U (1985) The cell wall. Prog Bot 47: 1-8

Kutschera U (1991) Regulation of cell expansion. In: Lloyd CW (ed) The cytoskeletal basis of plant growth and form. Academic Press, London, pp 149-158

Kutschera U, Bergfeld R, Schopfer P (1987) Cooperation of epidermal and internal tissues in auxin mediated growth of maize coleoptiles. Planta 170: 168-180

Kwon YW, Yim KO (1986) Paclobutrazol in rice. FFTC Book Ser Taiwan 35: 130-137

Lambert AM (1993) Microtubule-organizing centers in higher plants. Curr Opin Cell Biol 5: 116-122

Lang JM, Eisinger WR, Green, PB (1982) Effects of ethylene on the orientation of microtubules and cellulose microfibrils of pea epicotyls with polylamellate cell walls. Protoplasma 110: 5-14

Laskowski MJ (1990) Microtubule orientation in pea stem cells: a change in orientation follows the initiation of growth rate decline. Planta 181: 44-52

Laude HH, Pauli AW (1956) Influence of lodging on yield and other characters in winter wheat. Agron J 148: 453-455

Ledbetter MC, Porter KR (1963) A "microtubule" in plant cell fine structure. J Cell Biol 12: 239-250

Liu B, Joshi HC, Wilson TJ, Silflow CD, Palevitz BA, Snustad DP (1994) γ-Tubulin in *Arabidopsis*: gene sequence, immunoblot, and immunofluorescence studies. Plant Cell 6: 303-314

Lloyd CW, Seagull RW (1985) A new spring for plant cell biology: microtubules as dynamic helices. Trends Biochem Sci 10: 476-478

Lloyd CW, Slabas AR, Powell AJ, Lowe SB (1980) Microtubules, protoplasts and plant cell shape. An immunofluorescent study. Planta 147: 500-506

Lockhart J (1960) Intracellular mechanism of growth inhibition by radiant energy. Plant Physiol 35: 129-135

Luib M, Schott PE (1990) Einsatz von Bioregulatoren. In: Haug G, Schuhmann G, Fischbeck G (eds) Pflanzenproduktion im Wandel – Neue Aspekte in den Agrarwissenschaften. Verlag Chemie, Weinheim, pp 275-304

Makela P, Varaala L, Peltonensainio P (1996) Agronomic comparison of Minnesota-adapted dwarf oat with semi-dwarf, intermediate, and tall oat lines adapted to Northern growing conditions. Can J Plant Sci 76: 727-734

Marc J, Palevitz BA (1990) Regulation of the spatial order of cortical microtubules in developing guard cells in *Allium*. Planta 182: 626-634

Marc J, Sharkey DE, Durso NA, Zhang M, Cyr RJ (1996) Isolation of a 90-kDa microtubule-associated protein from tobacco membranes. Plant Cell 8: 2127-2138

Mcleod JG, Payne JF (1996) AC rifle winter rye. Can J Plant Sci 76: 143-144

Mita T, Katsumi M (1986) Gibberellin control of microtubule arrangement in the mesocotyl epidermal cells of the d_5 mutant of *Zea mays* L. Plant Cell Physiol 27: 651-659

Morejohn LC (1991) The molecular pharmacology of plant tubulin and microtubules. In: Lloyd CW (ed) The cytoskeletal basis of plant growth and form. Academic Press, London, pp 29-43

Nee M, Chiu L, Eisinger W (1978) Induction of swelling in pea internode tissue by ethylene. Plant Physiol 62: 902-906

Nick P (1997) Phototropic stimulation can shift the gradient of crown root emergence in maize. Bot. Acta 110: 291-297

Nick P, Furuya M (1992) Induction and fixation of polarity - early steps in plant morphogenesis. Dev Growth Differ 34: 115-125

Nick P, Furuya M (1993) Phytochrome-dependent decrease of gibberellin sensitivity. Plant Growth Regul 12: 195-206

Nick P, Furuya M (1996) Buder revisited: cell and organ polarity during phototropism. Plant Cell Environ 19: 1179-1187

Nick P, Schäfer E (1988) Spatial memory during the tropism of maize (*Zea mays* L.) coleoptiles. Planta 175: 380-388

Nick P, Schäfer E (1991) Induction of transverse polarity by blue light: an all-or-none response. Planta 185: 415-424

Nick P, Schäfer E (1994) Polarity induction versus phototropism in maize: auxin cannot replace blue light. Planta 195: 63-69

Nick P, Bergfeld R, Schäfer E, Schopfer P (1990) Unilateral reorientation of microtubules of the outer epidermal wall during photo- and gravitropic curvature of maize coleoptiles and sunflower hypocotyls. Planta 181: 162-168

Nick P, Furuya M, Schäfer E (1991) Do microtubules control growth in tropism? Experiments with maize coleoptiles. Plant Cell Physiol 32: 873-880

Nick P, Schäfer E, Furuya M (1992) Auxin redistribution during first positive phototropism in corn coleoptiles – microtubule reorientation and the Cholodny-Went theory. Plant Physiol 99: 1302-1308

Nick P, Yatou O, Furuya M, Lambert AM (1994) Auxin-dependent microtubule responses and seedling development are affected in a rice mutant resistant to EPC. Plant J 6: 651-663

Nick P, Lambert AM, Vantard M (1995) A microtubule-associated protein in maize is induced during phytochrome-dependent cell elongation. Plant J 8: 835-844

Nick P, Godbolé R, Wang QY (1997) Probing rice gravitropism with cytoskeletal drugs and cytoskeletal mutants. Biol Bull 192: 141-143

Nishiyama I (1986) Lodging of rice plants and countermeasure. FFTC Book Ser Taiwan 34: 152-163

Oda K, Suzuki M, Odagawa T (1966) Varietal analysis of physical characters in wheat and barley plants relating to lodging and lodging index. Bull Natl Inst Agric Sci Tokyo 15: 55-91

Parness J, Horwitz SB (1981) Taxol binds to polymerized tubulin in vitro. J Cell Biol 91: 479-487

Phinney BO, Spray CR (1990) Dwarf mutants of maize – research tools for the analysis of growth. In: Pharis RB, Rood SB (eds) Plant growth substances. Springer Verlag, Berlin Heidelberg New York, pp 65-73

Pinthus MJ (1973) Lodging in wheat, barley and oats: the phenomenon, its causes, and preventive measures. Adv Agron 25: 210-263

Robinson DG, Quader H (1981) Structure, synthesis, and orientation of microfibrils. IX: A freeze fracture investigation of the *Oocystis* plasma membrane after inhibitor treatments. Eur J Cell Biol 25: 278-288

Robinson DG, Quader H (1982) The microtubule-microfibril syndrome. In: Lloyd CW (ed) The cytoskeleton in plant growth and development. Academic Press, London, pp 109-126

Robson PRH, Mccormac AC, Irvine AS, Smith H (1996) Genetic engineering of harvest index in tobacco through overexpression of a phytochrome gene. Nat Biotechnol 14: 995-998

Sakiyama M, Shibaoka H (1990) Effects of abscisic acid on the orientation and cold stability of cortical microtubules in epicotyl cells of the dwarf pea. Protoplasma 157: 165-171

Sakiyama-Sogo M, Shibaoka H (1993) Gibberellin A₃ and abscisic acid cause the reorientation of cortical microtubules in epicotyl cells of the decapitated dwarf pea. Plant Cell Physiol 34: 431-437

Schlenker G (1937) Die Wuchsstoffe der Pflanzen. Lehmanns Verlag, München, pp 18-19

Schott PE, Lang H (1977) Mittel zur Regulierung des Pflanzenwachstums. BASF Patent No DE 27 55 940 C2, Deutsches Patentamt, München

Schott PE, Knittel H, Klapproth H (1984) Tetcyclacis – a new bioregulator for improving the development of young rice seedlings. ACS Symp Ser, Washington, USA, p 257

Schreiner C, Reed HS (1908) The toxic action of certain organic plant constituents. Bot Gaz USA 45: 73-102

Seagull RW (1990) The effects of microtubule and microfilament disrupting agents on cytoskeletal arrays and wall deposition in developing cotton fibers. Protoplasma 159: 44-59

Shibaoka H (1993) Regulation by gibberellins of the orientation of cortical microtubules in plant cells. Austr J Plant Physiol 20: 461-470

Smith H (1981) Adaptation to shade. In: Johnson CB (ed) Physiological processes limiting plant productivity. Butterworths, London, pp 159-173

Steeves TA, Sussex IM (1989) Patterns in plant development. Cambridge University Press, Cambridge

Stoppin V, Vantard M, Schmit AC, Lambert AM (1994) Isolated plant nuclei nucleate microtubule assembly: the nuclear surface in higher plants has centrosome-like activity. Plant Cell 6: 1099-1106

Sullivan KF (1988) Structure and utilization of tubulin isotypes. Annu Rev Cell Biol 1: 687-713

Tanimoto T, Nickell LG (1967) Re-evaluation of gibberellin for field use in Hawaii. Rep Haw Sugar Technol 1966, 184

Tolbert NE (1960) (2-Chloroethyl)trimethyl-ammonium chloride and related compounds as plant growth substances. II. Effect on growth of wheat. Plant Physiol 35: 380-385

Toyomasu T, Yamane H, Murofushi N, Nick P (1994) Phytochrome inhibits the effectiveness of gibberellins to induce cell elongation in rice. Planta 194: 256-263

Vaughan MA, Vaughn KC (1988) Carrot microtubules are dinitroaniline resistant. I. Cytological and cross-resistance studies. Weed Res 28: 73-83

Weibel RO, Pendleton JW (1961) Effect of artificial lodging on winter wheat grain yield and quality. Agron J 56: 187-188

Wymer CL, Lloyd CW (1996) Dynamic microtubules: Implications for cell wall patterns. Trends Plant Sci 1: 222-227

Wymer CL, Fisher DD, Moore RC, Cyr RJ (1996) Elucidating the mechanism of cortical microtubule reorientation in plant cells. Cell Motil Cytoskeleton 35: 162-173

Yuan M, Shaw PJ, Warn RM, Lloyd CW (1994) Dynamic reorientation of cortical microtubules from transverse to longitudinal in living plant cells. Proc Natl Acad Sci USA 91: 6050-6053

Zandomeni K, Schopfer P (1993) Reorientation of microtubules at the outer epidermal wall of maize coleoptiles by phytochrome, blue-light photoreceptor and auxin. Protoplasma 173: 103-112

Zandomeni K, Schopfer P (1994) Mechanosensory microtubule reorientation in the epidermis of maize coleoptiles subjected to bending stress. Protoplasma 182: 96-101

Ziegenspeck H (1948) Die Bedeutung des Feinbaus der pflanzlichen Zellwand für die physiologische Anatomie. Mikroskopie 3: 72-85

2 Control of plant shape

Peter Nick

Institut für Biologie II, Schänzlestr. 1, D-79104 Freiburg, Germany

2.1
Summary

In the past years, the establishment of body shape and Bauplan in animals has been shown to be based upon a spatiotemporal pattern in the activity of transcription factors. The discovery of homeotic genes controlling flower morphogenesis stimulated the view that plant morphogenesis might follow the same principles. Based on this idea, a mutant approach was initiated to screen for pattern deletion mutants in *Arabidopsis*. Interestingly, the characterization of these mutants lead not to the isolation of the anticipated transcription factors, but to genes that are related to the formation of the cell plate. This outcome moves the spatiotemporal control of cell division into the centre of interest. Plants have developed microtubular structures that seem to be intimately linked to the spatial control of cell division: radial microtubules, preprophase band, and phragmoplast. This chapter will begin with a brief discussion of the impact of plant shape and then discuss the control of division axis and symmetry in cell biological terms. This will be followed by a discussion of potential molecular mechanisms related to this spatial control. A second role of microtubules during the control of plant shape is connected to their participation in mechano and gravity sensing. The prospect part of the chapter will treat, in a speculative way, the possibilities to manipulate gravitropic set-point angles and thus the angle of side branches (e.g. to increase photosynthetic efficiency), but also the control of phyllotaxis (whorl pattern), root architecture and tuber formation (in potatoes).

2.2
The impact of plant shape and cell shape in agriculture

The control of plant height can be regarded as a special aspect in the larger context of shape control. In contrast to animals, where shape is fairly independent of the environment, plant shape does vary to a large extent. Morphogenetic plasticity has been the major evolutionary strategy of plants to cope with environmental changes, and fitness seems to be intimately linked to plant shape (Fig. 2.1). During evolution, mechanical stress has been the main factor that has shaped plant architecture. In aquatic plants, plant shape was maintained by buoyancy. This allowed amazing size to be reached on the basis of fairly simple architecture. The successful penetration into terrestrial habitats, however, required the formation of a flexible, but nevertheless resistant mechanical lattice. The development of this lattice, the vessel system, has been a central theme during the evolution of terrestrial plants. Thus, the evolution of land plants was to a large extent driven by the need to overcome the mechanical constraints of gravity that was not longer compensated by buoyancy. It appears that plant architecture is shaped by mechanical stress and there exists an influential school explaining plant morphogenesis in terms of mechanosensing (Goodwin and Trainor 1985; Selker et al. 1992).

The close relation between patterns of mechanical strain and plant shape is mirrored on the level of individual cells. As discussed in the previous chapter, cell shape depends on an interplay between the expanding protoplast driven by the swelling vacuole, and the cell wall as limiting and guiding counterforce. It is possible to understand the shape of individual cells in a plant tissue as manifestations of minimal mechanical tension (Thompson 1959). Conversely, if one succeeds in changing the shape of individual cells in a tissue, this should lead to a repartitioning in the patterns of mechanical strain and thus to altered resistance to exogenous mechanical stress.

Fig. 2.1. Importance of flexible morphogenesis for plant survival. This tree was hit by a blow from a volcanic erruption, but it was able to survive due to the ability to realign its axis of growth with gravity. The image was taken in the area of the volcano Ura-Bandai-san in the northern part of the Japanese main island Honshu.

The impact of shape control for agriculture is linked mainly to the resistance of crop plants to mechanical stress such as wind. The losses in yield that are caused by wind stress vary considerably between different crops, years and areas, but field trials comparing wind-sheltered with wind-exposed samples give estimates ranging between 20 and 50% for graminean crops and up to 80% for certain apple varieties (Grace 1977). In addition to plant length limiting lodging resistance (see Chap. 1), wind resistance is related to the general shape of a plant (Grace 1977). For instance, the resistance of root systems to uprooting by storms has been shown to depend on the angle between the primary root and the branch roots (Stokes et al. 1996). In addition to wind stress, changes in cell shape can influence the performance of a plant when it is challenged by other exogenous stresses. Xeromorphic species, for instance, are characterized by stomata that are deeply embedded in the tissue or by leaves that are covered by hairs or trichomes and these adaptations are expected to increase drought resistance as well as the resistance to UV irradiation. The agronomical relevance of shape control is not confined to increased stress resistance, however. The formation of harvestable products is commonly accompanied by dramatic morphogenetic events, such as the formation of tubers, bulbs or fruits. The manipulation of these morphogenetic events bears directly on marketable yield. Marketable yield is not necessarily the same as biomass. In order to compete on a saturated market it can be meaningful to partition the biomass into the production of fewer, but larger fruits or tubers. In other cases, as for instance in potato or tulip breeding, a higher number of smaller structures might be desired. The morphogenetic events involved in the formation of tubers or fruits might be utilized to optimize biomass partitioning between fruit number and fruit size. This approach does not require changes in source-sink relation or increases in photosynthetic efficiency. Plant shape can contribute, however, to photosynthetic efficiency as well – the increases in rice yield that were termed green revolution in Southeast Asia were rendered possible by changes in plant architecture. changes in the angle of the leaves allows the high-yield rice cultivars to intercept more light and thus to accumulate more energy. This architectural approach will be continued in the future to create rice plants where the panicles are situated lower than the leaves, preventing shading of the leaves by the panicles (Lee 1994).

2.3
Cellular mechanisms for shape control

Any attempt to manipulate plant shape or architecture must interfere with the shape of individual cells. In addition to the spatial control of cell expansion discussed in the previous chapter, it is cell division that contributes to cell shape. In a certain sense cell division has to be placed even upstream of cell expansion, because it defines the original cell axis and thus the framework in which expansion can take place. In some cases, cell division can even interfere with cell differentiation and thus with the developmental fate of a given cell.

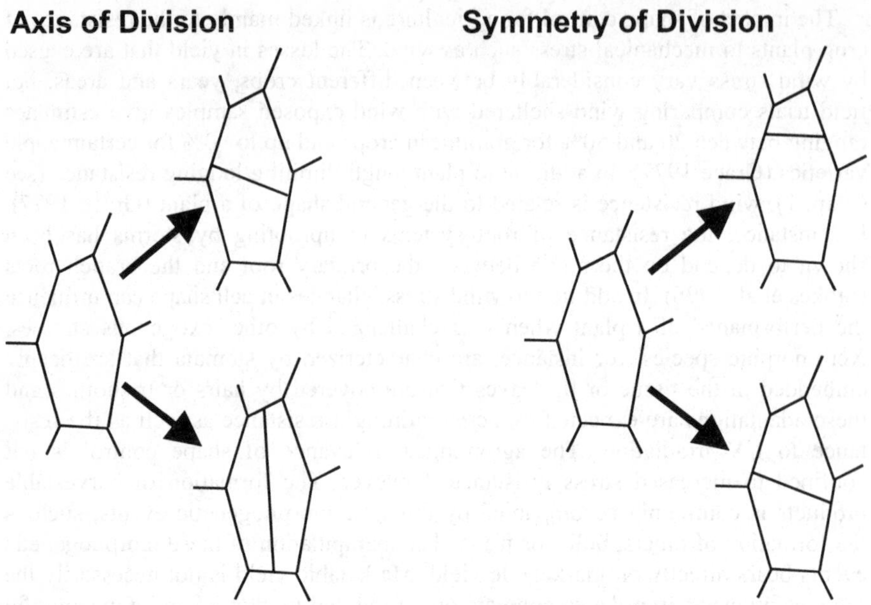

Axis of Division   Symmetry of Division

Fig. 2.2   Spatial control of axis and symmetry during the division of plant cells.

During cell division there are two aspects that are under tight spatial control: the axis of division and the symmetry of division (Fig. 2.2). Changes in axis or symmetry are relatively rare: in most cases, the cell axis is strictly maintained during division, and in most cases plant cells divide symmetrically. When such changes do occur, they are usually the earliest indication for dramatic morphogenetic responses that are in many cases induced by environmental signals. In fern protonemata that are cultivated in darkness or under red light, the division axis of the apical cell is aligned parallel to the main axis of the protonema. This axis is tilted by 90° in response to blue light and this switch of the division axis marks the transition towards two-dimensional growth and the formation of a prothallium (Mohr 1956; Wada and Furuya 1970). When the cell is returned to red light, the division axis will return to the original state and new protonemata form at the front edge of the prothallium. A similar realignment of cell division is observed during the wound response of higher plants, when the cells begin to divide perpendicularly to the wound surface (Hush et al. 1990).

Changes in division symmetry are often accompanied by cell differentiation, for instance during the formation of stomata (Bünning 1965) or during the formation of water-storage cells in the peat moss *Sphagnum* (Zepf 1952). In many spores and zygotes of lower plants, the first cell division is asymmetric and separates the prospective thallus from the prospective rhizoid (Quatrano 1978). When this first division is rendered symmetric by antimicrotubular drugs, cell differentiation is blocked, resulting in the formation of two thalli (Vogelmann et al. 1981). The first zygotic division of higher plants is asymmetric as well and, again, this asymmetry seems to influence cell differentiation. In the *gnom* mutant of *Arabidopsis thaliana* the first cell division is found to be symmetric, and this is accompanied by a dramatic change in the developmental fate of the daughter cells, leading to embryos with a defect apicobasal polarity (Mayer et al. 1993).

The *gnom* mutant had been selected in an attempt to screen for pattern deletion mutants (Mayer et al. 1991) in analogy to a similar approach in *Drosophila* (St. Johnston and Nüsslein-Volhard 1992). In animals, the control of the basic body pattern, the Bauplan, has been shown to depend on a spatiotemporal pattern in the activity of transcription factors, and a similar view has been stimulated by the discovery of homeotic genes that control flower morphogenesis (Weigel and Meyerowitz 1994). Interestingly, the molecular characterization of these mutants did not lead to the isolation of transcription factors as one might have expected from the results obtained in *Drosophila* or from flower morphogenesis. The cloning of the gene mutated in the *gnom* mutant revealed certain homologies to the yeast Sec7, a protein involved in the secretory pathway (Shevell et al. 1994). A second mutation, knolle, that was thought to participate in the separation of the epidermis from the inner tissues, was found to be related to syntaxin and appears to play a role in vesicle transport and the formation of the cell plate (Lukowitz et al. 1996). These observations emphasize the impact of spatially ordered cell division on plant morphogenesis.

The axis and symmetry of the cell division define the shape of the daughter cells and thus the preferential axis of cell expansion. It is possible, as discussed in Chapter 1, to switch the direction of cell expansion postmitotically by a reorientation of cortical microtubules. However, the ground state of the cell axis is laid down by the division process itself. Approaches to manipulate cell shape must therefore focus on the mechanisms that define the axis and the symmetry of cell division.

2.4
Preprophase band, phragmoplast and the spatial control of cell division

Higher plants have evolved specialized populations of microtubules that are not found in animals: cortical microtubules, the preprophase band (PPB) and the phragmoplast (Fig. 2.3). These plant-specific microtubule arrays participate in the spatial control of cell expansion and cell division. The cortical microtubules are characteristic for interphase cells and usually form parallel bundles perpendicular to the main axis of cell expansion (Fig. 2.3a). As pointed out in Chapter 1, they are involved in the directional control of cellulose deposition and in the axiality of cell growth. The present chapter will focus on those microtubular arrays that control axis and symmetry of cell division.

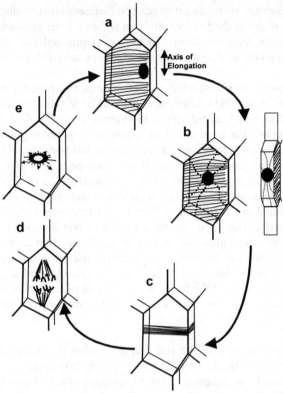

Fig. 2.3a-e. Microtubular arrays during the cell cycle of higher plants. **a** Elongating interphase cell with cortical microtubules. The nucleus is situated in the periphery of the cell. **b** Cell preparing for mitosis seen from above and from the side. The nucleus has moved towards the cell centre and is tethered by radial microtubules emanating from the nuclear envelope. **c** Preprophase band of microtubules. **d** Mitosis and division spindle. **e** Cell in telophase with phragmoplast that organizes the new cell plate and extends in centrifugal direction.

When a plant cell prepares for mitosis, this is heralded by a migration of the nucleus to the site, where the prospective cell plate will be formed. This movement is driven by the phragmosome, a specialized array of actin microfilaments (Katsuta and Shibaoka 1988; Lloyd 1991). At the same time, the cortical microtubules are gradually replaced by a new network of radial microtubules that emanate from the nuclear surface and often merge with the cortical cytoskeleton (Fig. 2.3b). The cortical microtubules subsequently disappear and are replaced almost instantaneously by a band of microtubules surrounding the cell equator, the preprophase band (Fig. 2.3c). This preprophase band is laid down in parallel to the direction of cortical microtubules and is tethered to the nucleus by the radial microtubules and by the phragmosome. The preprophase band marks site and direction of the prospective cell plate. Interestingly, it disappears with the formation of the division spindle that is usually organized in an axis perpendicular to the preprophase band and with the spindle equator situated in the plane marked by the preprophase band (Fig. 2.3d). When the daughter chromosomes have separated, a new array of microtubules, the phragmoplast, emerges at the site of the ensuing cell plate (Fig. 2.3e). The phragmoplast targets vesicle transport to the periphery of the growing cell plate. Electronmicroscopical evidence supports a model, where microtubules pull at tubular-vesicular outgrowths that emanate from the endoplasmatic reticulum (Samuels et al. 1995). The phragmoplast consists of a double ring of inter-digitating microtubules that grows in diameter with progressive extension of the cell plate. New microtubules are organized along the outer edge of the expanding phragmoplast (Vantard et al. 1990).

These observations assign a central role to nuclear migration for the spatial control of cell division. The nuclear migration seems to be driven by actin micro-filaments of the phragmosome (Lloyd 1991), because the movement is blocked by the actin inhibitor cytochalasin B (Katsuta and Shibaoka 1988). The radial micro-tubules, characteristic for cells where the nucleus has moved to the cell centre (Fig. 2.3b), support the actin cytoskeleton in the tethering of the nucleus to its new site. Upon treatment with antimicrotubular drugs such as colchicin (Thomas et al. 1977) or propyzamide (Katsuta and Shibaoka 1988), the nucleus can be loosened and displaced by mild centrifugation.

The formation of the preprophase band is initiated at the end of the S-phase and continues throughout the G2 phase (Gunning and Sammut 1990). The preprophase band predicts faithfully the symmetry and axis of the ensuing cell division. This is illustrated impressively during asymmetric cell divisions, for instance during stomatal development (Wick 1991) or in the response of root tissue to wounding (Hush et al. 1990). However, the preprophase band is discussed as being more than a true indicator for the spatial aspects of cell division.

The following evidence supports a causal relationship between the formation of the preprophase band and the control of division axis and symmetry:

1. In the *Arabidopsis* mutants *tonneau* and *fass* the preprophase band is lacking. The ordered pattern of cell divisions characteristic for the development of the wild type is replaced by a completely randomized pattern in both mutants (Traas et al. 1995; McClinton and Sung 1997).
2. In apical cells of fern protonemata, the formation of the preprophase band can be manipulated by either cold treatment (causing depolymerization of microtu-bules) or by centrifugation of the nucleus to the basal end of the cell (Murata and Wada 1991a). If the nucleus was displaced towards the cell base just prior to the formation of the preprophase band, the preprophase band was established in the new position of the nucleus at the cell base, and, subsequently, the new cell plate formed in the cell base as well. When the nucleus was centrifuged out of the apex somewhat later (when a preprophase band had already been formed), the nucleus

could induce a second preprophase band in its new position in the cell base. In these cells, the new cell plate was laid down randomly with respect to orientation and symmetry. These experiments elegantly demonstrate that the nucleus can induce and guide the formation of the preprophase band, and that the correct formation of the preprophase band is closely correlated to the correct deposition of the cell plate. These experiments seem to prove that the formation of preprophase band and the formation of the cell plate are linked directly. It should be mentioned, however, that, in meiotic cells, the division plane can be controlled in the absence of a preprophase band (Brown and Lemmon 1991).

These findings suggest that the preprophase band is the earliest manifestation of the division axis known so far. For the symmetry of division, however, nuclear migration and the nuclear envelope play the pivotal role. This conclusion is reached from experiments in fern protonemata, where the preprophase band had been eliminated by a cold shock causing microtubule depolymerization, and where the nucleus could organize a new preprophase band (Murata and Wada 1991b).

The establishment of the division spindle probably represents a bypass of the chain that links nuclear migration, formation of the preprophase band, and the induction of the phragmoplast: although the original axis of the spindle is always laid down in a direction perpendicular to the preprophase band, it can be tilted or distorted to oblique arrays due to space restrictions (Mineyuki et al. 1988). Surprisingly, this does not result in the formation of an oblique phragmoplast or an oblique cell plate, indicating that the spindle itself is uncoupled from the morphogenetic processes responsible for cell-plate formation. However, the spindle is the only microtubular array that could bridge the gap between the disappearance of the preprophase band (occurring at the onset of the metaphase) and the appearance of the phragmoplast (after the end of the anaphase). It remains enigmatic how the preprophase band can guide the formation of the phragmoplast when it disappears much earlier. The missing link might be the phragmosome, an actin structure that participates in nuclear migration and in the organization of the preprophase band and persists during mitosis (Lloyd 1991). The observations made in the *tonneau* and *fass* mutants (Traas et al. 1995; McClinton and Sung 1997) and the centrifugation experiments with fern protonemata (Murata and Wada 1991a,b) suggest that the guided organization of the phragmoplast requires the interaction of nuclear envelope, preprophase band and the phragmosome.

These considerations emphasize the central role of nuclear envelope, preprophase band, the network of radial microtubules and the phragmosome for the spatial control of cell division and thus for the control of cell shape. This conclusion raises the question of potential molecular targets to manipulate nuclear migration and the establishment of the radial microtubule/phragmosome network.

2.5
Potential molecular targets for cell-shape control

Premitotic nuclear migration can be disrupted by cytochalasin B and by antimicrotubular inhibitors (Thomas et al. 1977; Katsuta and Shibaoka 1988). This suggests that the radial microtubules appearing at that time cooperate with actin microfilaments during the movement and tethering of the nucleus in the prospective division plane. This step basically defines the symmetry of the ensuing cell division. The nuclear surface can then induce, in a second step, the formation of preprophase band and phragmosome in the plane of the prospective cell plate. This step defines the axis of the ensuing cell division. The decisive questions remain unanswered, however (Fig. 2.4a). How is the nuclear movement directed towards the prospective plane of division? How is the nuclear surface differentiated into an equatorial region, apparently capable of organizing the preprophase band, and two polar regions that seem to lack this ability.

Both processes seem to depend on the original polarity of the cell, as is illustrated by experiments with germinating fern spores (Vogelmann et al. 1981). In these spores, the first cell division is highly asymmetric and separates the thallus precursor from the rhizoid precursor. Prior to division, the primary nucleus undergoes a premitotic migration that can be blocked by antimicrotubular inhibitors. If this migration is blocked, a symmetric cell division is observed and both daughter cells give rise to thallus tissue. This result illustrates that the spore was highly polarized with respect to unknown developmental determinants and that these determinants had been uncompletely separated in consequence of the inhibitor treatment. This means, in other words, that the original cell polarity drives and directs the nuclear migration in untreated spores.

The nuclear envelope of higher plant cells has the intrinsic property to nucleate new microtubules (Lambert 1993), and the relation of microtubule nucleation versus microtubule elongation could be influenced by addition of mitotic cytosolic extracts to isolated plant nuclei (Stoppin et al. 1994). The nuclear envelope of premitotic plant cells contains γ-tubulin that is supposed to participate in the nucleation of microtubules (Liu et al. 1994) and CCT, a chaperone that specifically folds nascent tubulin, is located at the nuclear surface in interphase cells of maize (Himmelspach et al. 1997).

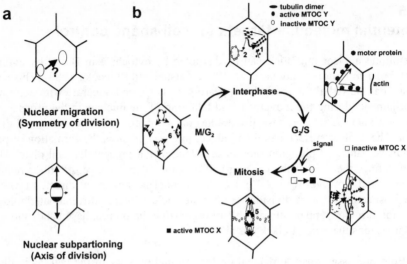

Fig. 2.4a,b. Possible mechanism in the control of division axis and symmetry. **a** Relation between symmetry and axis of cell division with nuclear migration and subpartitioning into different domains. **b** Potential role of cytoskeletal reorganization and division axis and symmetry. During interphase, tubulin dimers are partitioned into cortical microtubule arrays (**1**) due to the activity of cortical microtubule-organizing centers (MTOCs), whereas only few tubulin dimers are bound by the relatively inactive MTOCs that are situated on the nuclear envelope (**2**). In premitotic cells, the nucleus is moved to a central position by actin-based motors and tethered by microtubules (**7**). The final position is reached as force equilibrium proportional to the length of the actin bundle (and thus to the number of motors that can bind to this bundle). A different type of MTOCs (type X) is transported along actin to the nuclear poles and accumulates there (**4**). The MTOCs of type Y on nuclear surface and phragmosome are activated leading to a repartitioning of tubulin dimers towards the nucleus and the replacement of cortical microtubules by the preprophase band and radial microtubules that emanate from the nuclear envelope (**3**). At the onset of mitosis a different signal activates the MTOCs of type X that have accumulated at the cell pole, and inactivates the MTOCs of type Y. The MTOCs at the nuclear poles initiate the formation of spindle microtubules (**5**). During telophase, the MTOCs of type Y (localized at the newly established nuclear envelope and in the cell periphery) are reactivated and organize the phragmoplast and new cortical microtubules (**6**).

The organization of the preprophase band is accompanied by a phosphorylation of proteins. Some of these phosphorylated proteins are located at the nucleus (Young et al. 1994), whereas a cell-cycle dependent protein kinase, $p34^{cdc2}$, has been found to be colocalized with the preprophase band (Colasanti et al. 1993). The transition from cortical microtubules towards radial microtubules can be mimicked in interphase cells by cycloheximide, a blocker of protein synthesis (Mineyuki et al. 1994). This suggests that the radial array is a kind of default state, whereas the cortical array has to be actively maintained by the synthesis of proteins with a relatively short lifetime. Interestingly, the formation of a preprophase band was not inhibited by cycloheximide, indicating that the microtubules comprising the preprophase band are distinct in molecular terms from cortical microtubules. This conclusion is supported by experiments in wheat roots, where the formation of the preprophase band could be blocked by taxol, an inhibitor of microtubule depolymerization (Panteris et al. 1995). Although the preprophase band persisted in taxol-treated cells, a spindle could be formed, but it was multipolar and looked abnormal. The formation of the phragmoplast is inhibited in taxol treated cells (Yasuhara et al. 1993), suggesting that the depolymerization of the spindle is required to trigger the procession of the microtubular cycle. The proteins that participate in the establishment of the phragmoplast and the microtubule-driven transport of vesicles towards the growing edge of the cell plate are gradually being identified. Microtubule-associated proteins that bind dependent on the presence of ATP have been isolated from purified tobacco phragmoplasts (Yasuhara et al. 1992), and a plant dynamin-like protein, termed phragmoplastin, has been shown to be located across the whole width of the newly formed cell plate, suggesting that it might be associated with exocytotic vesicles that are depositing cell-plate material during cytokinesis (Gu and Verma 1995). The gene product encoded by the *knolle* gene, a member of the syntaxin family, had been identified via the corresponding *Arabidopsis* mutant and has been observed to be located in the phragmoplast as well (Lukowitz et al. 1996).

The molecular components controlling the spatial aspects of cell division are still far from being understood. Nevertheless, a first speculative model can be built already at this stage (Fig. 2.4b) that should open pathways for manipulation:

1. The cortical array of microtubules is actively maintained in interphase cells by proteins that have to be synthetized continuously (Mineyuki et al. 1994). If the activity of these proteins decreases, this will result in a rapid deterioration of the cortical array.
2. The nuclear envelope contains proteins that are able to nucleate new microtubules (Liu et al. 1994; Stoppin et al. 1994; Himmelspach et al. 1997), and it seems that this nucleating function is actively suppressed during interphase. This suppression requires protein synthesis (Mineyuki et al. 1994). In the simplest case, the suppression of microtubule nucleation at the nuclear envelope might be the direct consequence of elevated microtubule nucleation in the cortical plasma if both sites compete for a limited number of free tubulin dimers.
3. At the transition G2/S-phase this suppression is released (possibly by reduced microtubule nucleation in the cortical plasma with subsequent excess of tubulin dimers), and new microtubules form spontaneously at the nuclear envelope, giving rise to the radial network of microtubules.
4. In order to obtain directionality, the nucleation and elongation of radial microtubules must be guided by cell polarity (Vogelmann et al. 1981). As discussed in Chapter 1, the guiding principle might be the actin microfilaments that are aligned with the cell axis and connect the nucleus to the cell periphery throughout interphase. A component X that regulates microtubule nucleation (for instance a microtubule-associated protein) could be transported along actin towards the polar regions of the nucleus. By this mechanism, the nuclear envelope would be partitioned into different regions. In the equatorial region, where the concentration of X is expected be lower, cell-cycle dependent kinases could activate a different

factor, Y, that is bound to the actin microfilaments of the phragmosome and that triggers the nucleation of the preprophase band.

5. The activity of the nucleating factor Y that is located in the nuclear equator and in the preprophase band decreases at the same time as the nucleating factors of type X situated at the spindle poles become activated. The signal responsible for both changes could be identical (a mitotic cyclin, for instance), if it is an inhibitor of Y and simultaneously an activator of X.

6. At the end of anaphase, when the activity of this mitotic cyclin decreases, this will result in a shift of nucleation activity from X in favour of Y. New microtubules are nucleated in the prospective cell plate and establish the phragmoplast.

7. These events participate in the control of the division axis. The symmetry or asymmetry of division is defined by the initial nuclear migration driven by actin microfilaments. At first glance it may appear complex to tune the migration such that the nucleus will end up in the centre of the cell. A relatively simple model suggests that the actin cables that are typical for most interphase cells slide along each other with a sliding force that is roughly proportional to the length of the cable (for instance, because the number of actin motors that can bind to a long cable will be larger than for a shorter cable). By such a mechanism the nucleus would be pulled away from the cell periphery and should gradually reach the centre of the cell, where the length of the connecting cables would come to equilibrium. In the case of an asymmetric division the equilibrium could be shifted by an intracellular gradient in the concentration of actin motors.

2.6
Role of microtubules in gravity- and mechanosensing

The impact of microtubules on plant shape is not confined to the spatial aspects of cell division. They participate, in addition, in the sensing of mechanical stimuli such as gravity and this opens a second pathway to manipulate plant shape via the microtubular cytoskeleton.

Efficient links between gravity perception and morphogenesis are vital for survival. In plants, this link becomes manifest in two basic phenomena:

1. When the orientation of a plant is changed with respect to gravity it will respond by a very sensitive bending response that restores the original orientation (gravitropism).

2. The formation and orientation of new organs is often adjusted with respect to gravity (gravimorphosis).

Both events seem to involve the microtubular cytoskeleton: microtubules are endowed with axiality and high flexural rigidity (Gittes et al. 1993), ideal properties for the amplification of weak mechanic stresses. The potential energy of the amyloplasts that commonly serve as statoliths in the gravity-sensing process of higher plants barely exceeds thermal noise (Björkman 1988). This implies the existence of highly efficient signal amplification systems participating in the early steps of gravity sensing in plants. In fact, a number of observations indicate that microtubules might be involved in this amplification mechanism:

1. Tubulin polymerization and depolymerization respond directly to mechanical fields, causing an alignment of microtubules with respect to the gravity vector even during microtubule assembly in vitro (Tabony and Job 1992).
2. A mechanosensitive calcium channel in onion cells becomes irreversibly inhibited when the membranes are pretreated with the antimicrotubular drug ethyl-N-phenylcarbamate (Ding and Pickard 1993).
3. Gravitropism that is triggered by the pressure of the sedimenting amyloplasts (Kuznetsov and Hasenstein 1996) upon mechanosensitive ion channels can be blocked by antimicrotubular drugs in the *Chara* rhizoid (Friedrich and Hertel 1973), in moss protonemata (Schwuchow et al. 1990; Walker and Sack 1990) and in coleoptiles of maize (Nick et al. 1991) and rice (Nick et al. 1997) at concentrations that leave phototropism and/or growth essentially unaltered.
4. Rice mutants where microtubules are less dynamic (Nick et al. 1994), or wild type coleoptiles where microtubules have been rendered less dynamic by treatment with taxol exhibit a strong delay in the gravitropic response (Nick et al. 1997).
5. The reorientation of cortical microtubules that accompanies the gravitropic response in coleoptiles (Nick et al. 1991) is blocked by taxol along with the gravitropic bending (Nick et al. 1997).
6. The application of mechanical fields (Hush and Overall 1991), high pressure (Cleary and Hardham 1993) or artificial bending of the tissue (Zandomeni and Schopfer 1994) can induce a reorientation of cortical microtubules from an originally transverse to a longitudinal array.

The reorientation of cortical microtubules in response to gravitropic stimulation has been observed in both coleoptiles (Nick et al. 1991) and roots (Blancaflor and Hasenstein 1993). In maize coleoptiles, the microtubules in the epidermal cells of the upper flank of the stimulated organ assumed a longitudinal orientation, whereas the microtubules in the lower flank reinforced their original transverse orientation. The time course of this response was consistent with a model where gravitropic stimulation induced a lateral shift of auxin transport towards the lower organ flank and, consequently, a depletion of auxin in the upper flank. The microtubular response was thought to be primarily triggered by this decrease in auxin concentration rather than by gravity itself. In maize roots, however, where a similar reorientation could be observed in the cortex (Blancaflor and Hasenstein 1993), the time course of reorientation suggested that microtubules respond to the changes in growth rate induced by gravity rather than to the gravitropic stimulus directly.

This poses the question whether a direct gravitropic response of microtubules can be demonstrated at all. In moss protonemata, the microtubules that are adjacent to the amyloplasts were redistributed in response to gravitropic stimulation (Schwuchow et al. 1990). In rice coleoptiles, the microtubules in the gravity-sensitive cells of the bundle sheath are found to reorient rapidly from transverse to longitudinal. This reorientation can be blocked by taxol, leading to a delay of the gravitropic response (Nick et al. 1997). On the other hand, if microtubule reorientation is artificially promoted by elimination of actin microfilaments with cytochalasin D, the onset of gravitropic bending is accelerated (Nick et al. 1997). These correlations suggest that dynamic microtubules amplify and accelerate the sensing and early transduction of gravity. This conclusion is supported by the altered organization of the cytoskeleton that is observed specifically in the statocytes of root caps where the endoplasmic microtubules are depleted from the central parts of the cell and where the rate of amyloplast sedimentation can be elevated by chemically induced disintegration of cortical microtubules (Baluška et al. 1997).

The impact of gravimorphosis is illustrated by the simple observation that roots form at the basal pole of a plant. Although a considerable amount of phenomenological work was dedicated to this problem at the turn of the century (Vöchting 1878; Sachs 1880; Goebel 1908), the underlying mechanisms have remained obscure so far. One reason for this problem has been certainly the use of adult organs, where polarity has already been fixed and is hard to invert. In the past years, new systems have been introduced that may be more appropriate for the study of gravimorphosis. Germinating fern spores initiate their development with a first asymmetric division that separates a larger, vacuolated rhizoid precursor from a smaller and denser thallus rhizoid precursor. This first cell division appears to be of formative character: when it is rendered symmetric by treatment with antimicrotubular drugs (Vogelmann et al. 1981) the two daughter cells give rise both to thalloid tissue. The axis of this first division is strictly aligned with gravity. When the spore is tilted subsequent to the first division, the rhizoid grows in the wrong direction and cannot correct this error (Edwards and Roux 1994). Prior to division, at the time when the spore is competent to the aligning influence of gravity, a vivid migration of the nucleus towards the lower half of the spore is observed. This movement is not a simple sedimentation process because it is rhythmic and sometimes exhibits short periods of active sign reversal, indicating that the nucleus is tethered to the cell wall (Edwards and Roux 1997). The action of antimicrotubular compounds strongly suggests that this guiding mechanism is based on microtubules that must align with respect to the gravity vector. It should be mentioned that a similar mechanism of gravimorphosis has been detected during the determination of the dorsiventral axis in frog eggs (Gerhart et al. 1981), where the axis is determined by an interplay of gravity-dependent sedimentation of yolk particles, sperm-induced nucleation of microtubules and self-amplifying alignment of newly formed microtubules that drive cortical rotation (Elinson and Rowning 1988).

This example demonstrates again that microtubule-dependent nuclear migration is a prerequisite for the spatial control of cell division and the formative response to environmental stimuli.

2.7
Manipulation of gravitropic set-point angles

Gravitropic sensitivity is not a constant, but varies depending on development, light regime and type of organ. This variability, although often neglected, must be regarded as an essential element of the gravitropic response and contributes to the characteristic shape of a given plant. The angles between side branches and main axis of a tree are usually observed to decrease with height such that the lower branches are oriented almost horizontally, whereas the younger branches near the top of the tree grow almost vertically (Fig. 2.5). This developmental pattern of branching angles is characteristic for each species and allows optimal utilization of solar irradiation. Since the older branches are longer and heavier, one might be tempted to assume that they are simply bent to a stronger degree. However, the horizontal growth habit is maintained actively, as can be shown when the organ is brought into a vertical position. The current position is apparently compared to an internal value that changes with development and has been termed gravitropic setpoint angle (Myers et al. 1994). The gravitropic setpoint angle of a root is opposed to that of a main shoot, it is perpendicular for organs that grow plagiotropically and it has been shown to depend on light, developmental stage and organ (Myers et al. 1994). The characteristic shape of a dicotyledonous seedling, for instance, can be shown to depend on a switch in the gravitropic setpoint angle with increasing maturation of the cells along the length of the hypocotyl (Myers et al. 1994).

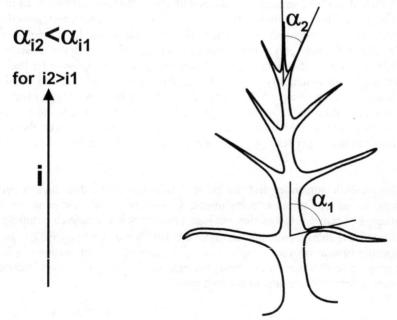

Fig. 2.5. Relation of the gravitropic setpoint angle to the shape of a tree. The angle is larger in the basal older parts of the tree and decreases towards the top of the tree.

The mechanism defining the gravitropic setpoint angle of a given organ has remained as fascinating as it is enigmatic, but it is clear that it cannot originate from differences in the sensing process itself: amyloplasts sediment downwards in the shoot as well as in the root, but this sedimentation is interpreted in just the opposite way. The direction of the gravity vector must somehow be compared to an internal vector that can switch depending on development and cell type. A switch of the internal vector has been reported to be induced artificially by high concentrations of auxin (Harrison and Pickard 1989) or by treatment with lithium (Millet and Pickard 1988), both treatments that are known to interfere with cell polarity. If cell polarity can be altered in a part of the cells, this should result in a changed gravitropic setpoint angle. A second target to manipulate the setpoint angle is the threshold of gravitropic sensitivity. The potential stimulus conveyed by the sedimentation of amyloplasts barely exceeds the threshold set by thermal noise. The primary signal must be either amplified or accumulated by temporal integration. Both possibilities imply the participation of a second partner that enhances the sensitivity of signal perception. The observations listed above suggest that this second partner is the cytoskeleton. Microtubules could thus be used to reduce the sensitivity threshold that can be sensed and by this to increase the maximal angle that is tolerated without triggering a counteracting growth response.

If microtubules are rendered less dynamic by either drugs such as taxol (Nick et al. 1997) or by respective mutations (Nick et al. 1994), this causes a delayed gravitropic response in rice coleoptiles (Nick et al. 1997). The bending response itself, once it has been initiated, proceeds with about the same efficiency as in the control plants. This indicates that the response system (auxin transport, growth) as such has been left essentially unaffected. The step responsible for the delay must be situated in the early part of the transduction chain defining gravitropic sensitivity. When plants of the mutant *ER31*, where microtubules appear to be less dynamic, are grown to maturity, a marked change in the gravitropic behaviour of the leaf sheaths can be observed (Fig. 2.6). The leaf sheaths of wild-type plants are almost erect with a slight increase in the angle between leaf sheath and the main axis of the plant when the leaves become older. In the mutant, however, this angle is conspicuously larger resulting in an almost fan-like appearance of the leaf system.

This result demonstrates that manipulation of microtubular dynamics might be used to change gravitropic setpoint angles. As tools one might use genes for microtubule-associated proteins (for instance proteins that can induce bundling of microtubules such as the elongation factor EF-1α (Durso and Cyr 1994) or genes for specific tubulin isotypes. For the purpose of agricultural application it will be necessary in most cases to combine this approach with precise developmental fine-tuning for the expression of the host genes.

Fig. 2.6. Increased gravitropic setpoint angle in leaf sheaths of the rice mutant *ER31*, where the dynamics of microtubules is reduced.

To increase photosynthetic efficiency of rice, one might conceive, for instance, a system where the leaf sheath remains erect, but the blade of the leaf is oriented in horizontal position to intercept more light. For this purpose, a microtubule-bundling protein could be expressed preferentially in the base of the leaf blade such that the reduced turnover of microtubules will cause a reduced gravitropic sensitivity of the cells lining the border between sheath and blade. This is expected to increase the tolerance for a more horizontal blade orientation. In the ideal case, the manipulation should be restricted to those cells where gravity is sensed to minimize any potential side effects. The existence of specific tubulin isotypes that participate in sensory amplification of mechanosensation in *Caenorhabditis* (Huang et al. 1995) opens the possibility that statocyte-specific promotors might be found in plants as well.

Gravitropic setpoint angles are intimately linked to the very heart of plant architecture. If they are manipulated by changing microtubular dynamics in the gravity-sensing cells this would open pathways to control the core principles of plant shape.

2.8
Manipulation of phyllotaxis

The formation of new leaf primordia in the apical meristem follows a self perpetuating pattern termed phyllotaxis. One of the first indications of primordia initiation is a reorientation of cortical microtubules in a ring of epidermal cells around the margin of the prospective primordium. Except for the changed behaviour of microtubules, these cells are indistinguishable from their neighbours with respect to dimension or growth axis (Hardham et al. 1980). The orientation of microtubules can differ between neighbouring cells to a dramatic extent.

However, microtubules become increasingly aligned with time producing a smooth transition of microtubule orientation between neighbouring cells. The shift in microtubule orientation causes a shift in the axis of cell expansion leading to a bulging of the apical meristem at the site of primordia formation. This is then followed by a shift in the axis of cell division and the outgrowth of a new primordium.

Phyllotaxis depends primarily on the agent that induces some cells to be committed for microtubule reorientation. It is possible to predict the site where this orientation will occur by calculating the positions of minimal energy based on the pattern of mechanical stress produced by the older preexisting primordia (Green 1992). The overall size of the apex in relation to the region that is partitioned to a primordium is expected to be important for the resulting pattern (for instance, distichous versus spiral phyllotaxis). The gradual alignment of microtubules seems to be based on their ability to sense and respond to wall stresses.

To test the minimal-stress hypothesis experimentally, beads that had been loaded with purified expansin were loaded on the apical meristem of tomato plants (Fleming et al. 1992). Expansin can interrupt the hydrogen bonds between cellulose fibres and thus induce loosening of cell walls (McQueen-Mason et al. 1992). The localized release of cell-wall tension by the beads caused a bulging of the apical meristem and in some cases the formation of ectopic leaf-like structures (Fleming et al. 1992). Interestingly, these ectopic leaves produced a reversal in the sign of phyllotaxis indicating that, in fact, the pattern of wall stress had been disturbed such that the positions of minimal energy moved to different positions.

From this experiment one can conceive two ways to manipulate phyllotaxis: Change either the extensibility of the cell wall or the microtubular response to mechanic stimuli. The first might be hampered by the fact that successful overexpression of appropriated genes is not sufficient. The corresponding proteins would have to be exported correctly through the Golgi system into the cell wall across several membrane boundaries. This is supposed to raise certain difficulties that could be circumvented by the second approach. For the microtubular response to wall stress it is necessary that the microtubules are interconnected with elements of the cell wall across the membrane (Williamson 1991). If this interaction is loosened, this should result in a reduced reorientation; if it is tightened, the microtubules should reorient more swiftly in a larger area of the apex.

Candidates for such microtubule-cell-wall linker proteins have been described in higher plants (Nick et al. 1995; Marc et al. 1996). If they are cloned and overexpressed in the apex, a larger proportion of the apex is expected to be partitioned to the primordium and this would favour distichous phyllotaxis over spiral phyllotaxis due to space limits. In extreme cases the apical meristem would be gradually "used up" because more cells are partitioned to leaf formation than can be regenerated by the meristem. This would result in a limited number of leaf organs, i.e. determinate growth. If the link between cell wall and microtubules is weakened in the respective antisense approach, fewer cells would be partitioned to the prospective primordium. In distichous species this should result in a shift of phyllotaxis to a spiral arrangement.

Spiral phyllotaxis allows the plant to maximize the amount of sunlight intercepted by the leaves. Under arid conditions, where light is not limiting, a more compact structure of the plant could be advantageous and distichous phyllotaxis might be the desired trait. In a refined version of this approach, the expression of the microtubule-cell wall linkers might be put under the control of a mechanosensitive promoter, which should allow precise fields of microtubule alignment to be produced and thus more complex phyllotactic patterns.

2.9
Manipulation of root architecture

Plant roots have the function of anchoring the plant firmly in the ground and providing a maximum surface where the uptake of water and nutrients and the interaction with soil microorganisms can take place. From these tasks three targets for the manipulation of root architecture can be derived:

1. Increase in rooting depth. The more deeply the branching part of the rooting system is embedded in the soil, the more resistant it will be against uprooting (Stokes et al. 1996). Elongation of roots is usually limited by either drought or lowered oxygen content (Gianì and Breviario 1996) occurring, for instance, in flooded or condensed soils. This block of elongation is correlated to changes in the expression of β-tubulin isotypes (Gianì and Breviario 1996).
2. Optimized branching angles of side roots. Optimal resistance to uprooting is achieved by side roots that branch from the main root at an angle of 90° and by second-order side roots that branch from the side roots at an angle of 20° (Stokes et al. 1996). The branching angles depend on a switch in the gravitropic setpoint angle during the formation of side roots and second-order side roots. This gravitropic setpoint angle is expected to depend on the dynamics of microtubules (see above).
3. Stimulation of root-hair formation and elongation. The formation of root hairs depends on an asymmetric division giving rise to the protodermal trichoblast (Peterson and Farquhar 1996). The elongation of root hairs occurs by tip growth and is driven by acropetal vesicle transport driven by actin microfilaments and longitudinal cables of microtubules. The structure of the root hairs contributes to the so-called rhizosheath, a soil sheath that protects the root surface from desiccation and hosts beneficial soil-borne microorganisms.

To manipulate these targets, the following approaches could be designed:

1. The elongation of rice roots is accompanied by a high level of expression of β tubulins and when root elongation is blocked by anoxia, the level of one isotype of β-tubulin (OS-TUB16) is decreased (Gianì and Breviario 1996). A similar de-

crease is observed in the rice coleoptile, but there the loss of OS-TUB16 is compensated for by increased expression of other β-tubulin isotypes, ensuring a high level for the whole population of β-tubulin transcripts and a high rate of elongation even under anoxia (Gianì and Breviario 1996). These data suggest that the inhibition of *OS-TUB16* transcription by anoxia limits root elongation, a phenomenon of great agronomical importance, because the soil under rice paddies is characterized by low concentrations of oxygen limiting the development of sturdy root systems. To overcome this problem one might envisage a construct where a high level of β-tubulin transcripts is maintained by transgenic plants where the *OS-TUB16* is controlled by a root-specific promotor that is constitutively active. A more subtle approach would search for the anoxia-stimulated isotypes of β-tubulin that compensate for the inhibition of *OS-TUB16* in the coleoptile (Gianì and Breviario 1996). The promotor of these isotypes could then be complemented by sequences that confer expression in the root. This approach is expected to transfer the coleoptile behaviour of β-tubulin transcription to the root, but only upon anoxia. In the presence of oxygen the transgenic plant would not be distinct from a wild-type plant.

2. To change the branching patterns of side roots, one could try to modify gravitropic setpoint angles by changing the dynamics of tubulin assembly and disassembly (see above). However, to make this approach work, it is necessary to find promotors that are specifically active in lateral roots. A second target might be those areas of the root meristem where the direction of microtubular arrays is variable and where the root can respond to morphogenetic signals that allows root architecture to be tuned with the conditions in the soil (Barlow and Parker 1996). These variable areas are those sites where morphogenetic decisions are taken and where the plane of cell divisions is flexible. If the direction of the preprophase band could be altered in these cells (see above), this might influence the axis of prospective side roots.

3. The formation of root hairs involves asymmetric cell divisions producing specialized epidermal cells, the so-called trichoblasts. This asymmetric division involves a shift in the positioning of the preprophase band (Peterson and Farquhar 1996). Although one could design approaches where this apparently formative cell division is manipulated via the preprophase band, it is probably easier to make use of genes that code for transcription factors and seem to suppress the differentiation of all epidermal cells into trichoblasts such as *transparent testa glabra* in *Arabidopsis* (Galway et al. 1994). The elongation of root hairs occurs by tip growth. This is illustrated by the observation that root hairs of the tip1 mutant, where pollen tube growth is impaired, are short and branched in a way that is characteristic for trichomes of *Arabidopsis* (Schiefelbein et al. 1993). Root hairs are characterized by longitudinal bundles of microtubules that are thought to participate in the directional transport of vesicles towards the growing tip of the root hair. The importance of the microtubular cytoskeleton for tip growth is emphasized by the analysis of the *zwichel* mutant of *Arabidopsis*, where the extension of trichomes is impaired. The cloning of the mutated gene by T-DNA tagging led to the identification of a kinesin-like protein (Oppenheimer et al.

1997). If the length of tip-growing cells is limited by the availability of such microtubule-driven motors, it should be possible to stimulate root hair elongation and thus the extension of the rhizosheet by overexpression of kinesins in root hairs. A good candidate for such motor proteins seems to be the gene affected in the *tip1* mutant (Schiefelbein et al. 1993).

2.10
Manipulation of tuber formation

The formation of tubers and bulbs is accompanied by a block of cell elongation and the initiation of lateral swelling. A tuber-inducing factor that had been isolated from potato leaves from plants that were rendered competent for tuber formation by cultivation in short days (Koda et al. 1988) was identified as a glucoside of jasmonic acid (Yoshihara et al. 1989). Jasmonates seem to be involved in the formation of tubers and bulbs in other species as well such as yam, Jerusalem artichoke, sweet potato, onion or garlic (Koda 1997).

These morphogenetic events are accompanied by dramatic responses of the cortical microtubules: the formation of onion bulbs is accompanied by the disruption of cortical microtubules (Mita and Shibaoka 1984), whereas the formation of potato tubers seems to involve a reorientation of cortical microtubules into longitudinal arrays (Fujino et al. 1995). Both responses have the effect that the original reinforcement of elongation by a transverse orientation of microtubules and microfibrils is lost. The reorientation of cortical microtubules in lateral potato shoots can be induced by the tuber-inducing factor jasmonic acid (Koda 1997). Gibberellins that suppress tuber formation inhibit this reorientation of cortical microtubules (Sanz et al. 1996). Additionally, gibberellins suppress in a similar way the disruption of cortical microtubules that accompanies bulb formation in onion (Mita and Shibaoka 1984). These findings suggest that the reorientation of cortical microtubules (or, alternatively, the disruption of cortical microtubules) is necessary for tuber formation. The inducing trigger seems to be jasmonic acid, whereas gibberellins seem to play an antagonistic role. The microtubule response appears to be not sufficient, however. Jasmonates that are administered to the main stem cannot induce the formation of tubers, although they can induce the reorientation of microtubules and the termination of cell elongation (Koda 1997). This indicates that the competence for tuber formation must involve a second unknown factor. This factor is responsible for the maintenance of cell expansion, once cell elongation has come to an end due to the jasmonic acid-triggered reorientation of cortical microtubules.

Stimulated by these observations the following potential approaches might be developed:

1. The number of tubers might be elevated, if microtubule reorientation or disruption is induced in additional sites that are principally competent for tuber formation. This would result in a higher number of tubers, probably correlated to a

smaller size of individual tubers. This approach could be interesting for potato breeders wishing to obtain a higher number of seed potatoes from a limited number of parental plants (this situation is relevant during the propagation of transgenic lines). To induce microtubule reorientation, one could use the natural trigger, jasmonic acid, although the problem of undesired side effects should be carefully checked. Alternatively, one could try anticytoskeletal drugs that can induce swelling of roots (Baskin and Bivens 1995).The problem of these approaches is their global mode of action. To stimulate the number of tubers, it would be more appropriate to induce microtubular responses at specific sites on the shoot epidermis. For this purpose, a microtubule-disrupting factor or a factor triggering microtubule reorientation could be expressed under control of a promotor that is active in defined subsets of epidermal cells. Disruption of microtubules can be triggered via the calcium-calmodulin pathway (Fisher et al. 1996). One might, for instance, conceive a system where a stomata-specific promotor is coupled to genes that cause an overflow of calcium. The level of calcium should then be elevated in a certain area around individual stomata and initiate lateral swelling in a limited area. By reducing the activity of the respective promotor one might achieve that this pathway is activated only in a limited number of stomata, possibly yielding defined number of tubers.

2. The number of tubers could be lowered with a concomitant increase in the size of individual tubers. For this approach it might be sufficient to stimulate lateral swelling globally by increasing the activity of the trigger (jasmonate) or by application of microtubule-disrupting drugs at low concentration. The stimulation of swelling is possibly sufficient to suppress the initiation of tuber formation in other sites simply by mutual competition of tuber-formating sites for a limiting flow of assimilates.

The manipulation of plant shape via the microtubular cytoskeleton is certainly still far ahead and the approaches that are envisaged in the outlook section of this chapter are undoubtedly still speculative in character. Nevertheless, their potential for plant morphogenesis is large. The biotechnological approach of the future must meet several criteria: it must be efficient, is must have minimal side effects and it should allow for a control of amplitude. The essential idea of the potential approaches presented in this chapter is to utilize the morphogenetic responses that occur naturally in the plant. These responses are either targeted to cells where they usually do not occur or they are minimally modified to obtain changes in response amplitude. In the ideal case, manipulation of plant shape should be brought about without the help of heterologous genes or promotors. It might be sufficient to recombine endogenous genes and promotors in an intelligent way. The strong impact of the microtubular cytoskeleton for morphogenesis makes us expect that relatively subtle changes in assembly dynamics or microtubule bundling can produce conspicuous and nevertheless specific effects on plant shape.

References

Baluška F, Kreibaum A, Vitha S, Parker JS, Barlow PW, Sievers A (1997) Central root cap cells are depleted of endoplasmic microtubules and actin microfilament bundles – implications for their role as gravity-sensing statocytes. Protoplasma 196: 212-223

Barlow PW, Parker JS (1996) Microtubular cytoskeleton and root morphogenesis. Plant Soil 187: 23-36

Baskin TI, Bivens NJ (1995) Stimulation of radial expansion in *Arabidopsis* roots by inhibitors of actomyosin and vesicle secretion, but not by various inhibitors of metabolism. Planta 197: 514-521

Björkman T (1988) Perception of gravity by plants. Adv Bot Res 15: 1-4

Blancaflor EB, Hasenstein KH (1993) Organization of cortical microtubules in graviresponding maize roots. Planta 191: 230-237

Brown RC, Lemmon BE (1991) The cytokinetic apparatus in meiosis: control of division plane in the absence of a preprophase band of microtubules. In: Lloyd CW (ed) The cytoskeletal basis of plant growth and form. Academic Press, London, pp 259-273

Bünning E (1965) Die Entstehung von Mustern in der Entwicklung von Pflanzen. In: Ruhland W (ed) Handbuch der Pflanzenphysiologie vol 15/1, Springer, Berlin Heidelberg New York, pp 383-408

Cleary AL, Hardham AR (1993) Pressure induced reorientation of cortical microtubules in epidermal cells of *Lolium rigidum* Leaves. Plant Cell Physiol 34: 1003-1008

Colasanti J, Cho SO, Wick S, Sundaresan V (1993) Localization of the functional p34^{cdc2} Homolog of Maize in Root Tip and Stomatal Complex Cells: Association with Predicted Division Sites. Plant Cell 5: 1101-1111

Ding JP, Pickard BG (1993) Mechanosensory calcium-selective cation channels in epidermal cells. Plant J 3: 83-110

Durso NA, Cyr RJ (1994) A calmodulin-sensitive interaction between microtubules and a higher plant homolog of elongation factor 1α. Plant Cell 6: 893-905

Edwards ES, Roux SJ (1994) Limited period of graviresponsiveness in germinating spores of *Ceratopteris richardii*. Planta 195: 150-152

Edwards ES, Roux SJ (1997) The influence of gravity and light on developmental polarity of single cells of *Ceratopteris richardii* gametophytes. Biol Bull 192: 139-140

Elinson RB, Rowning B (1988) Transient array of parallel microtubules in frog eggs: potential tracks for a cytoplasmic rotation that specifies the dorsoventral axis. Develop Biol 128: 185-197

Fisher DD, Gilroy S, Cyr RJ (1996) Evidence for opposing effects of calmodulin on cortical microtubules. Plant Physiol 112: 1079-1087

Fleming AJ, McQueen-Mason, Mandel T, Kuhlemeier C (1992) Induction of leaf primordia by the cell wall protein expansin. Science 276: 1415-1418

Friedrich U, Hertel R (1973) Abhängigkeit der geotropischen Krümmung von der Zentrifugalbeschleunigung. Z Pflanzenphysiol 70: 173-184

Fujino K, Koda Y, Kikuta Y (1995) Reorientation of cortical microtubules in the sub-apical region during tuberization in single-node stem segments of potato in culture. Plant Cell Physiol 36: 891-895

Galway ME, Masucci JD, Lloyd AM, Walbot V, Davis RW, Schiefelbein JW (1994) The *TTG* gene is required to specify epidermal cell fate and cell patterning in the *Arabidopsis* root. Dev Biol 166: 740-754

Gerhart J, Ubbeles G, Black S, Hera K, Kirschner M (1981) A reinvestigation of the role of the grey crescent in axis formation in *Xenopus laevis*. Nature 292: 511-516

Giani S, Breviario D (1996) Rice β-tubulin mRNA levels are modulated during flower development and in response to external stimuli. Plant Sci 116: 147-157

Gittes F, Mickey B, Nettleton J, Howard J (1993) Flexural rigidity of microtubules and actin filaments measured from thermal fluctuations in shape. J Cell Biol 120: 923-934

Goebel K (1908) Einleitung in die experimentelle Morphologie der Pflanzen. Teubner, Leipzig, pp 218-251

Goodwin BC, Trainor LEH (1985) Tip and whorl morphogenesis in *Acetabularia* by calcium-regulated strain fields. J Theor Biol 117: 79-106

Grace J (1977) Plant response to wind. Academic Press, London, pp. 121-134,135-142

Green PB (1992) Cellulose orientation in primary growth: an energy-level model for cytoskeletal alignment. Curr Top Plant Biochem Physiol 11: 99-117

Gu XJ, Verma DPS (1995) Phragmoplastin, a dynamin-like protein associated with cell plate formation in plants. EMBO J 15: 695-704

Gunning BES, Sammut M (1990) Rearrangements of microtubules involved in establishing cell division planes start immediately after dna synthesis and are completed just before mitosis. Plant Cell 2: 1273-1282

Hardham AR, Green PB, Lang JM (1980) Reorganization of cortical microtubules and cellulose deposition during leaf formation of *Graptopetalum paraguayense*. Planta 149: 181-195

Harrison M, Pickard BG (1989) Auxin asymmetry during gravitropism by tomato hypocotyls. Plant Physiol 89: 652-657

Himmelspach R, Nick P, Schäfer E, Ehmann B (1997) Developmental and light-dependent changes of the cytosolic chaperonin containing TCP-1 (CCT) subunits in maize seedlings, and the localization in coleoptiles. Plant J 12: 1299-1310

Huang M, Gu G, Ferguson EL, Chalfie M (1995) A stomatin-like protein necessary for mechanosensation in *C. elegans*. Nature 378: 292-295

Hush JM, Overall RL (1991) Electrical and mechanical fields orient cortical microtubules in higher plant tissues. Cell Biol Int Rev 15: 551-560

Hush JM, Hawes CR, Overall RL (1990) Interphase microtubule re-orientation predicts a new cell polarity in wounded pea roots. J Cell Sci 96: 47-61

Katsuta J, Shibaoka H (1988) The roles of the cytoskeleton and the cell wall in nuclear positioning in tobacco BY-2 cells. Plant Cell Physiol 29: 403-413

Koda Y (1997) Possible involvement of jasmonates in various morphogenetic events. Physiol Plant 100: 639-646

Koda Y, Omer ESA, Yoshihara T, Shibata H, Sakamura S, Okazawa Y (1988) Isolation of a specific tuber-inducing substance from potato leaves. Plant Cell Physiol 29: 1047-1051

Kuznetsov OA, Hasenstein KH (1996) Intracellular magnetophoresis of amyloplasts and induction of root curvature. Planta 198: 87-94

Lambert AM (1993) Microtubule-organizing centers in higher plants. Curr Opin Cell Biol 5: 116-122

Lee B (1994) Filling the world's rice bowl. IRRI, Los Baños

Liu B, Joshi HC, Wilson TJ, Silflow CD, Palevitz BA, Snustad DP (1994) γ-Tubulin in *Arabidopsis*: gene sequence, immunoblot, and immunofluorescence studies. Plant Cell 6: 303-314

Lloyd CW (1991) Cytoskeletal elements of the phragmosome establish the division plane in vacuolated plant cells. In: Lloyd CW (ed) The cytoskeletal basis of plant growth and form. Academic Press, London, pp 245-257

Lukowitz W, Mayer U, Jürgens G (1996) Cytokinesis in the *Arabidopsis* embryo involves the syntaxin-related KNOLLE gene product. Cell 84: 61-71

Marc J, Sharkey DE, Durso NA, Zhang M, Cyr RJ (1996) Isolation of a 90-kDa microtubule-associated protein from tobacco membranes. Plant Cell 8: 2127-2138

Mayer U, Torres Ruiz RA, Berleth T, Miséra S, Jürgens G (1991) Mutations affecting body organization in *Arabidopsis* embryo. Nature 353: 402-407

Mayer U, Büttner G, Jürgens G (1993) Apical-basal pattern formation in the *Arabidopsis* embryo: studies on the role of the *gnom* gene. Development 117: 149-162

McClinton RS, Sung ZR (1997) Organization of cortical microtubules at the plasma membrane in *Arabidopsis*. Planta 201: 252-260

McQueen-Mason S, Durachko DM, Cosgrove DJ (1992) Endogenous proteins that induce cell wall expansion in plants. Plant Cell 4: 1425-1433

Millet B, Pickard BG (1988) Early wrong way response occurs in orthogravitropic maize roots treated with lithium. Physiol Plant 72: 555-559

Mineyuki Y, Iida H, Anraku Y (1994) Loss of microtubules in the interphase cells of onion (*Allium cepa* l.) root tips from the cell cortex and their appearance in the cytoplasm after treatment with cycloheximide. Plant Physiol 104: 281-284

Mineyuki Y, Marc J, Palevitz BA (1988) Formation of the oblique spindle in dividing guard mother cells of *Allium*. Protoplasma 147: 200-203

Mita T, Shibaoka H (1984) Gibberellin stabilizes microtubules in onion leaf sheath cells. Protoplasma 119: 100-109

Mohr H (1956) Die Abhängigkeit des Protonemenwachstums und der Protonemapolarität bei Farnen vom Licht. Planta 47: 127-158

Murata T, Wada M (1991a) Effects of centrifugation on preprophase-band formation in *Adiantum* protonemata. Planta 183: 391-398

Murata T, Wada M (1991b) Re-formation of the preprophase band after cold-induced depolymerization of microtubules in *Adiantum* protonemata. Plant Cell Physiol 32: 1145-1151

Myers AB, Firn RD, Digby J (1994) Gravitropic sign reversal – a fundamental feature of the gravitropic perception or response mechanisms in some plant organs. J Exp Bot 45: 77-83

Nick P, Schäfer E, Hertel R, Furuya M (1991) On the putative role of microtubules in gravitropism of maize coleoptiles. Plant Cell Physiol 32: 873-880

Nick P, Yatou O, Furuya M, Lambert AM (1994) Auxin-dependent microtubule responses and seedling development are affected in a rice mutant resistant to EPC. Plant J 6: 651-663

Nick P, Lambert AM, Vantard M (1995) A microtubule-associated protein in maize is induced during phytochrome-dependent cell elongation. Plant J 8: 835-844

Nick P, Godbolé R, Wang QY (1997) Probing rice gravitropism with cytoskeletal drugs and cytoskeletal mutants. Biol Bull 192: 141-143

Oppenheimer DG, Pollock MA, Vacik J, Szymanski DB, Ericson B, Feldmann K, Marks MD (1997) Essential role of a kinesin-like protein in *Arabidopsis* trichome morphogenesis. Proc Natl Acad Sci USA 94: 6261-6266

Panteris E, Apostolakos P, Galatis B (1995) The effect of taxol on *Triticum* preprophase root cells – preprophase microtubule band organization seems to depend on new microtubule assembly. Protoplasma 186: 72-78

Peterson RL, Farquhar ML (1996) Root hairs – specialized tubular cells extending root surfaces. Bot Rev 62: 1-40

Quatrano RS (1978) Development of cell polarity. Annu Rev Plant Physiol 29: 489-510

Sachs J (1880) Stoff und Form der Pflanzenorgane. Arb Bot Inst Würzburg 2: 469-479

Samuels AL, Giddings TH, Staehelin LA (1995) Cytokinesis in tobacco BY-2 and root tip cells - a new model of cell plate formation in higher plants. J Cell Biol 130: 1345-1357

Sanz MJ, Mingocastel A, Vanlammeren AAM, Vreugdenhil D (1996) Changes in the microtubular cytoskeleton preceed in vitro tuber formation in potato. Protoplasma 191: 46-54

Schiefelbein J, Galway M, Masucci J, Ford S (1993) Pollen tube and root-hair tip growth is disrupted in a mutant of *Arabidopsis thaliana*. Plant Physiol 103: 979-985

Schwuchow J, Sack FD, Hartmann E (1990) Microtubule distribution in gravitropic protonemata of the moss *Ceratodon*. Protoplasma 159: 60-69

Selker JML, Steucek GL, Green PB (1992) Biophysical mechanisms for morphogenetic progressions at the shoot apex. Dev Biol 153: 29-43

Shevell DE, Leu WM, Gilmor CS, Xia GX, Feldmann KA, Chua NH (1994) EMB30 is essential for normal cell division, cell expansion, and cell adhesion in *Arabidopsis* and encodes a protein that has similarity to Sec7. Cell 77: 1051-1062

St.Johnston D, Nüsslein-Volhard C (1992) The origin of pattern and polarity in the *Drosophila* embryo. Cell 68: 201-219

Stokes A, Ball J, Fitter AH, Brain P, Coutts MP (1996) An experimental investigation of the resistance of model root systems to uprooting. Ann Bot 78: 415-421

Stoppin V, Vantard M, Schmit AC, Lambert AM (1994) Isolated plant nuclei nucleate microtubule assembly: the nucleus surface in higher plants has centrosome-like activity. Plant Cell 6: 1099-1106

Tabony J, Job D (1992) Gravitational symmetry breaking in microtubular dissipative structures. Proc Natl Acad Sci USA 89: 6948-6952

Thomas DDS, Dunn DM, Seagull RW (1977) Rapid cytoplasmic responses of oat coleoptiles to cytochalasin B, auxin, and colchicine. Can J Bot 55: 1797-1800

Thompson DW (1959) On growth and form. University Press, Cambridge, pp 465-644

Traas J, Bellini C, Nacry P, Kronenberger J, Bouchez D, Caboche M (1995) Normal differentiation patterns in plants lacking microtubular preprophase bands. Nature 375: 676-677

Vantard M, Levilliers N, Hill AM, Adoutte A, Lambert AM (1990) Incorporation of *Paramecium* axonemal tubulin into higher plant cells reveals functional sites of microtubule assembly. Proc Natl Acad Sci USA 87: 8825-8829

Vöchting H (1878) Über Organbildung im Pflanzenreich. Cohen, Bonn

Vogelmann TC, Bassel AR, Miller JH (1981) Effects of microtubule-inhibitors on nuclear migration and rhizoid formation in germinating fern spores (*Onoclea sensibilis*). Protoplasma 109: 295-316

Wada M, Furuya M (1970) Photocontrol of the orientation of cell division in *Adiantum*. I. Effects of the dark and red periods in the apical cell of gametophytes. Dev Growth Differ 12: 109-118

Walker LM, Sack FD (1990) Amyloplasts as possible statoliths in gravitropic protonemata of the moss *Ceratodon purpureus*. Planta 181: 71-77

Weigel D, Meyerowitz EM (1994) The ABCs of floral homeotic genes. Cell 78: 203-290

Wick SM (1991) The preprophase band. In: Lloyd CW (ed) The cytoskeletal basis of plant growth and form. Academic Press, London, pp 231-244

Williamson RE (1991) Orientation of cortical microtubules in interphase plant cells. Internat Rev Cytol 129: 135-206

Yasuhara H, Sonobe S, Shibaoka H (1992) ATP-sensitive binding to microtubules of polypeptides extracted from isolated phragmoplasts of tobacco BY-2 cells. Plant Cell Physiol 33: 601-608

Yasuhara H, Sonobe S, Shibaoka H (1993) Effects of taxol on the development of the cell plate and of the phragmoplast in tobacco BY-2 cells. Plant Cell Physiol 34: 21-29

Yoshihara T, Omer ESA, Koshino H, Sakamura S, Kikuta Y, Koda Y (1989) Structure of a tuber inducing stimulus from potato leaves (*Solanum tuberosum* L.). Agric Biol Chem 53: 2835-2837

Young T, Hyams JS, Lloyd CW (1994) Increased cell-cycle-dependent staining of plant cells by the antibody MPM-2 correlates with preprophase band formation. Plant J 5: 279-284

Zandomeni K, Schopfer P (1994) Mechanosensory microtubule reorientation in the epidermis of maize coleoptiles subjected to bending stress. Protoplasma 182: 96-101

Zepf E (1952) Über die Differenzierung des *Sphagnum*blatts. Z Bot 40: 87-118

3 Control of Wood Structure

Ryo Funada
Department of Forest Science, Hokkaido University, Sapporo 060-8589, Japan

3.1
Summary

The secondary xylem cells of woody plants, such as tracheids and wood fibres, have cell walls with a highly organized structure. The orientation of cellulose microfibrils in the primary wall determines the direction of cell elongation and expansion, thereby controlling the shape and size of xylem cells. In contrast, the orientation of microfibrils in the secondary wall, in particular in the thick middle layer (S_2 layer), is closely related to several of the mechanical properties of wood. Thus, the ability to control the orientation of cellulose microfibrils in the secondary wall might allow us to change the quality of wood and its products. There is considerable evidence that the dynamics of cortical microtubules are closely related to the orientation and localization of newly deposited cellulose microfibrils in the differentiating tracheids or wood fibres. Thus, it seems very likely that manipulation of cortical microtubules would allow control of the orientation of microfibrils, with a consequent improvement of wood quality.

3.2
Significance of microfibril orientation for wood structure

Wood, which is a renewable resource, has been used for millennia as a raw material. It is used currently for lumber, furniture, pulp and paper, chemicals and fuels. Wood is produced by the vascular cambium of living organisms, namely trees. Moreover, wood quality within a single species can vary markedly, depending on environmental factors such as climate, soil conditions and the spacing of growing trees, all of which affect the cambial growth of trees. In addition, wood quality varies among species and within individual trees according to cambial age, stem position and distance from the crown. Tree-to-tree variability within members of a species under identical growth conditions shows that genetic factors also influence wood quality. These observations indicate the strong possibility that wood quality might be improved and unified not only by silvicultural treatments, such as pruning, thinning and fertilization, but also by breeding to select genetically desirable trees (Panshin and de Zeeuw 1980; Zobel and van Buijtenen 1989; Zobel and Jett 1995). Recently developed molecular biological approaches also have the potential to improve wood quality, in particular the chemical composition of wood components such as lignin (Higuchi 1997). Differences in wood quality, in particular of mechanical properties, are largely due to differences in wood structure. Thus, wood structure is one of most important targets in attempts to control wood quality.

The orientation of microfibrils, referred to as the microfibril angle (the angle between cellulose microfibrils and the main cell axis) in the secondary wall is one of the most important ultrastructural characteristics that determine the properties of wood (Cave and Walker 1994). In particular, the angles of the middle layer (S_2 layer), which is the thickest layer in the secondary wall, are of major importance. The microfibril angles of the S_2 layer are negatively correlated with the modulus of elasticity (MOE) of wood (Wardrop 1951; Cave 1968). Thus, wood with large microfibril angles has low strength. The microfibril angles of the S_2 layer can be used as parameters for the assessment of the MOE in logs (Hirakawa et al. 1997). In addition, microfibril angles in the S_2 layer affect the shrinkage and swelling of wood in response to changes in moisture content (Barber and Meylan 1964; Harris and Meylan 1965). Usually, wood exhibits limited longitudinal shrinkage because of the steep orientation of cellulose microfibrils in the S_2 layer. However, wood where the S_2 layer is characterized by large microfibril angles, such as juvenile wood or compression wood, exhibits greater longitudinal shrinkage (up to 10%; Tsoumis 1991). Large changes in dimensions in consequence of shrinkage result in changes of shape, warping and collapse in wood and its products. These observations indicate that a biological approach to the control of microfibril angles in the secondary wall, in particular in the S_2 layer, might provide a powerful tool for changing the properties of wood. This Chap. describes the involvement, in secondary xylem cells, of cortical microtubules in the control of the direction of cell expansion and the structure of the cell wall, in particular the microfibril angles.

3.3
Cellular mechanisms of wood formation

In spite of the great economic importance of wood, the processes of cytodifferentiation during its formation are not yet fully understood (Catesson 1994; Higuchi 1997). By contrast, our understanding of xylem differentiation in vitro is quite advanced (Aloni 1987; Fukuda 1996). One of the major reasons for this difference is technical difficulties originating from the work with highly vacuolated cambial cells and their derivatives that are easily damaged during sampling (Catesson 1974, 1990; Goosen de Roo and van Spronsen 1978; Chaffey et al. 1997c). For example, for investigations of cambial cells at the ultrastructural level, different fixatives have to be used for different tissues or even for the same tissue sampled during different times of the year (Kidwi and Robards 1969; Catesson 1974). Therefore, the detailed information on wood formation needed for the development of biotechnological approaches to the control of wood structure is still missing.

Increases in the diameter of tree stems are due to the activity of the vascular cambium (Catesson 1994; Larson 1994). The cambium is defined as the actively dividing layer of cells that lies between, and gives rise to, the secondary xylem and phloem (IAWA Committee 1964). The cambium consists of fusiform initials and ray initials. The mean lengths of cambial fusiform initials range from 1100 to 4,000 µm in conifers and from 170 to 940 µm in hardwoods (Larson 1994). The lengths vary depending on the species and the age of the cambium. The periclinal division of cambial cells leads to an increase in stem diameter. Growth originating from cambial activity is designated secondary growth. The periclinal division of cambial cells produces the secondary phloem on the outside and the secondary xylem on the inside. However, the amount of secondary xylem produced is usually much higher than the amount of secondary phloem (Larson 1994). In vigorous trees, the increase in the number of xylem cells is ten times greater than for phloem cells (Imagawa 1981). Thus, matured xylem cells account for the most part of wood.

The stages in the development of xylem cells can be categorized as follows: cambial cell division, cell enlargement, cell wall thickening, lignification and cell death (Panshin and de Zeeuw 1980; Thomas 1991). The periclinal division of cambial cells in the tangential-longitudinal plane results in two daughter cells. As soon as a cell loses the ability to divide, it will start to differentiate. Cambial fusiform cells differentiate into tracheids, vessel elements, wood fibres and axial parenchyma cells, while ray initial cells differentiate into ray parenchyma cells and ray tracheids. Cells derived from cambial fusiform cells increase in length and diameter as they approach their final shape during differentiation. For example, tracheids in conifers increase only slightly in length but increase considerably in radial diameter (Fig. 3.1).

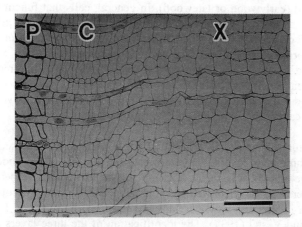

Fig. 3.1. Light micrograph of a cross-section of differentiating tracheids in *Abies sachalinensis*. The radial diameter of the differentiating tracheids increases gradually from the cambium towards the xylem. P Phloëm; C cambium; X xylem. **Bar** 100 µm.

By contrast, vessel elements do not increase significantly in length, whereas their increase in radial and tangential diameter is conspicuous. The very thin (often less than 0.1 μm thick) and plastic cell wall that is characteristic for the stage of cell enlargement is called the primary wall. Cellulose, a linear polymer composed of D-glucose with β-1,4-glycoside linkage, is the major component of the cell wall. Cellulose is highly crystalline and has very high tensile strength. Thus, cellulose microfibrils form a framework in the cell wall. The primary wall consists of loose aggregates of cellulose microfibrils (Harada 1965; Harada and Côté 1985). This structure allows relatively unimpeded expansion of the cells derived from the cambium. In addition, cytochemical observations have shown that the primary wall of cambial cells has a characteristic chemical composition that allows for considerable extensibility (Catesson 1990, 1994; Baïer et al. 1994; Catesson et al. 1994; Guglielmino et al. 1997). The activities of certain enzymes that are involved in the metabolism of cambial cell walls change at the onset of xylem differentiation (Catesson 1994; Guglielmino et al. 1997). Localized loosening of the cell wall, in particular at cell junctions, also occurs during cell expansion and the partial lysis of the cell wall is closely associated with a localized decrease in the level of calcium ions that are bound to the cell wall (Rao 1985; Funada and Catesson 1991; Baïer et al. 1994; Catesson et al. 1994).

When cell expansion is almost complete, well-ordered cellulose microfibrils are deposited on the inner surface of the primary wall, establishing the so called secondary wall (Harada 1965). Once the formation of the secondary wall has begun, no further expansion of cells occurs. Continuous deposition of the secondary wall increases the thickness of the cell wall. The thickness of the cell wall varies depending on cell function, cambial age and the season at which the cell is formed (earlywood or latewood). In general, cells that function to support the tree, such as tracheids and wood fibres, form thick secondary walls. Thus, the ultrastructure of tracheids and wood fibres is of great importance in defining the mechanical properties of wood. The cell wall supports the heavy weight of the tree itself and functions in the transport of water to the top of the tree, which can sometimes reach more than 100 m in height. In addition, the cell wall prevents microbial and insect attack, thereby protecting the tree during its very long life, that in some cases can exceed several thousands of years.

The structure of the secondary wall is not homogeneous. The secondary wall consists of three main layers: the outermost layer (S_1 layer), the middle layer (S_2 layer) and the innermost layer (S_3 layer) facing the lumen side, respectively. Each of these layers can be identified by polarizing and light microscopy following staining with iodine, and they were already described by Kerr and Bailey (1934) and Bailey and Vestal (1937). The identification of the three layers is facilitated by differences in orientation of the cellulose microfibrils in the various layers. The cellulose microfibrils in the S_1 and S_3 layers form a flat helix relative to the cell axis, whereas those in the S_2 layer form a steep helix. Detailed models for the structures of cell walls in tracheids and wood fibres have been proposed from the results of electron microscopy (Wardrop 1964; Harada and Côté 1985). Electron-microscopic observations of cross-sections reveal that, in general, the S_2 layer dominates. About 80% of the cell wall (in terms of thickness) is occupied by the S_2 layer in tracheids and wood fibres. Thus, the microfibril angles of the secondary walls in these cells, as determined by polarizing microscopy or X-ray diffraction analysis (Cave 1966; Yamamoto et al. 1993), are largely determined by the orientation of cellulose microfibrils in the S_2 layer.

With the onset of secondary wall deposition, the lignification begins at the intercellular layer, progressing to the primary wall and eventually to the secondary wall (Takabe et al. 1981b; Terashima et al. 1988). When lignification has been completed, cell death (cell autolysis) occurs immediately in most xylem cells.

Xylem cells that contribute to the mechanical support of the tree or to the conduction of water pass through the developmental stages mentioned above to become mature cells. However, cells that function in storage, such as ray parenchyma cells, remain alive for several years without immediate autolysis of cell organelles.

3.4
The interaction of microtubules and microfibrils in the primary wall

The turgor pressure (the pressure of the protoplast against the cell wall) within cells originates from the vacuole and provides the driving force for the enlargement of cells in plants. The increase in the volume of the vacuole is derived from a gradient in the water potential between cytoplasm and vacuole and the apoplast (Kutschera 1991). When the turgor pressure in the cell exceeds the yield point of the cell wall, the cell can expand. As the cell expands, the cell wall becomes stiffer and, consequently, its yield point increases. Then the rate at which the cell expands decreases gradually and finally expansion ceases.

The turgor pressure is exerted equally in all directions within a cell. If there were no reinforcement mechanism, cells might be expected to expand spherically. However, the cellulose microfibrils, with their considerable tensile strength, reinforce the cell wall and resist expansion in response to the turgor pressure. The predominant orientation of cellulose microfibrils in the cell wall is usually perpendicular to the direction of cell expansion (Green 1980; Taiz 1984).

In elongating cells, cellulose microfibrils are oriented transversely (Green and Kings 1966). The change in the direction of growth is related to the reorientation of cellulose microfibrils (Ridge 1973; Lang et al. 1982). It is generally accepted that the orientation of newly deposited cellulose microfibrils on the inner surface of the primary wall determines the direction and extent of cell expansion, thereby determining the final shape and size of the cell.

When transverse sections of secondary xylem cells are examined by polarizing microscopy, the primary wall is only slightly birefringent as a result of the differences in orientation of the cellulose microfibrils in the primary wall. Electronmicroscopic observations, mainly with samples prepared by replica methods, have shown that the cellulose microfibrils in the primary wall are not well ordered and are separated from each other by relatively large interspaces (Harada and Côté 1985). Randomly oriented cellulose microfibrils are observed during cell expansion of tracheids, wood fibres, and ray parenchyma cells (Imamura et al. 1972; Fujii et al. 1977; Thomas 1991; Abe et al. 1995b). The loose, random network of cellulose microfibrils in the primary wall allows for easy expansion.

Wardrop (1958) observed that, in macerated tracheids of *Pinus radiata*, the cellulose microfibrils were oriented approximately axially on the outer surface, whereas they were transverse relative to the cell axis on the inner surface. Observations by optical microscopy, X-ray diffraction, and electronmicroscopy have yielded diagrammatic representations for the organization of the primary wall of a typical tracheid or wood fibre (Wardrop 1964; Wardrop and Harada 1965; Harada and Côté 1985). In such models, the thin primary wall is composed of two layers with differently oriented cellulose microfibrils. Cellulose microfibrils observed by field emission-scanning electronmicroscopy (FE-SEM) are shown in Fig. 3.2. FE-SEM provides images of relatively large areas at high resolution. It is a particularly useful tool to follow changes in the orientation of newly deposited cellulose microfibrils during xylem differentiation.

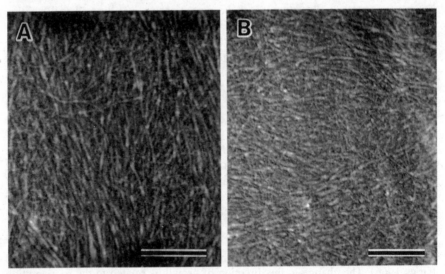

Fig. 3.2A,B. Electron micrographs (FE-SEM) of cellulose microfibrils (viewed from the lumen side) on the inner surface of the primary wall in differentiating tracheids of *Abies sachalinensis*. The axes of the tracheids in the photographs are oriented vertically. **A** At the first stage of cell expansion. **B** At the final stage of cell expansion. **Bar** 0.5 μm.

Since all tracheids or wood fibres derived from cambial fusiform initials are aligned in a radial direction, successive aspects of the formation of cell walls can be observed in a radial file within a single specimen. Cambial derivatives are a suitable system to follow the process of differentiation of secondary xylem in situ. The orientation of the cellulose microfibrils of the radial walls in tracheids of conifers changes during cell expansion (Fig. 3.2; Abe et al. 1995b). The cellulose microfibrils on the innermost surface of the primary wall are not well-ordered. Most of cellulose microfibrils in the tracheids at the early stage of cell expansion are predominantly oriented longitudinally. Longitudinally oriented cellulose microfibrils might act to restrain the turgor-driven longitudinal expansion to the cell axis. As the cell expands, the orientation of cellulose microfibrils on the innermost surface changes from longitudinal to transverse. At the final stage of cell expansion, cellulose microfibrils are oriented transversely to the cell axis. The tracheids elongate by only 5-15% but they expand radially by 200-500% during differentiation (Bailey 1920). Thus, in the cambial derivatives of fusiform initials the primary wall serves first to facilitate lateral expansion. In addition, transversely oriented cellulose microfibrils at the final stage of cell expansion probably prevent further lateral expansion. Therefore, the mechanism that controls the orientation of cellulose microfibrils in the primary wall determines the shape and the size of xylem cells.

Observations in a wide variety of plant cells have revealed that cortical microtubules play an important role in the orientation of newly deposited cellulose microfibrils (Ledbetter and Porter 1963; Hepler and Palevitz 1974; Gunning and Hardham 1982; Robinson and Quader 1982; Giddings and Staehelin 1991; Seagull 1991; Shibaoka 1994). Cellulose is synthesized by enzyme complexes that are often referred to as terminal complexes and that are localized in the plasma membrane (Herth 1985; Schneider and Herth 1986; Itoh 1991). It has been postulated that cortical microtubules that are closely associated with the plasma membrane guide the movement of these complexes. Coalignment of cortical microtubules and cellulose microfibrils has been observed in the cells of lower and higher plants. In addition, microtubule-depolymerizing agents, such as colchicine, usually disrupt the orientation of cellulose microfibrils (Robinson et al. 1976; Srivastava et al. 1977; Hogetsu and Shibaoka 1978). These observations support the hypothesis that cortical microtubules control the orientation of newly deposited cellulose microfibrils. Two models have been proposed for the mechanism by which cortical microtubules control the orientation of microfibrils (Heath and Seagull 1982; Gidding and Staehelin 1991). In model 1, cellulose synthase complexes "ride" directly on cortical microtubules, with putative physical links between microtubules and the cellulose-synthase complexes (Heath 1974). The linked molecules could be pulled by dynein-like motor proteins. In model 2, cellulose-synthase complexes move in membrane channels delimited by cortical microtubules. The complexes might be propelled forward by forces that result from the polymerization and crystallization of cellulose microfibrils. Gidding and Staehelin (1988) favoured model 2 because they found cellulose synthase complexes between or adjacent to cortical microtubules, rather than directly on top of cortical microtubules. By contrast, Vesk et al. (1996) recently observed, by high-resolution SEM, that individual cortical microtubules were positioned directly adjacent to individual cellulose microfibrils in the cell wall, suggesting the possibility that the orientation of each microfibril might be controlled directly by cortical microtubules.

The average diameter of microtubules is small, about 24 nm. Thus, the arrangement of cortical microtubules has been observed mainly by transmission electronmicroscopy (TEM) since the first observations by Ledbetter and Porter (1963). However, it is difficult to observe microtubules over a wide region of plant materials by TEM. With the successful introduction of indirect immuno-fluorescence microscopy, it became possible to visualize microtubules over large areas within plant tissues (Lloyd 1987; Lloyd et al. 1979; Wick et al. 1981). In addition, confocal laser scanning microscopy subsequent to immunofluorescence staining or microinjection of a fluorescence analogue made it possible to construct three-dimensional images of microtubules in plant cells (Yuan et al. 1994, 1995; Abe et al. 1995a; Lloyd et al. 1996; Funada et al. 1997; Furusawa et al. 1998). This method provides a powerful tool to follow the dynamics of microtubules (Fig. 3.3).

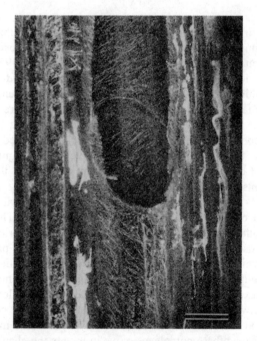

Fig. 3.3. An immunofluorescence projection image, obtained by confocal laser scanning microscopy, showing the arrangement of cortical microtubules in differentiating vessel elements of *Populus maxi-mowiczii*. **Bar** 25 μm.

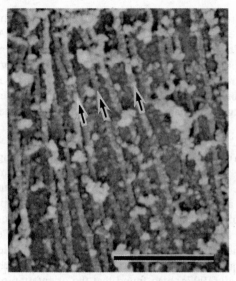

Fig. 3.4. Electron micrograph (FE-SEM) of cortical microtubules (**arrows**; viewed from the lumen side) in differentiating tracheids of *Abies sachalinensis*. Cortical microtubules are oriented in a steep Z-helix during formation of the middle layer (S_2 layer) of the secondary wall. **Bar** 0.5 µm.

In addition, SEM at higher resolution than conventional SEM also allows investigation of cytoplasmic components in wide regions of plant tissues. FE-SEM has also been used to observe the fine structure of the cytoskeleton in plant cells (Fig. 3.4; Hirakawa 1984; Abe et al. 1994; Vesk et al. 1994; 1996, Prodhan et al. 1995a).

The presence of cortical microtubules in cambial cells and their derivatives in woody plants has been confirmed by TEM (Cronshaw 1965; Srivastava 1966; Srivastava and O'Brien 1966; Robards and Humpherson 1967; Itoh 1971; Nobuchi and Fujita 1972; Fujita et al. 1974, 1978; Tsuda 1975; Barnett 1981; Côté and Hanna 1985; Inomata et al. 1992; Prodhan et al. 1995a; Chaffey et al. 1997a,b,d), SEM (Hirakawa 1984; Abe et al. 1994; Prodhan et al. 1995a) and immuno-fluorescence microscopy (Uehara and Hogetsu 1993; Abe et al. 1995a, b; Prodhan et al. 1995a; Chaffey et al. 1997a,b,c; Funada et al. 1997; Furusawa et al. 1998).

In developing wood fibres, cortical microtubules appear to be joined to the plasmalemma by bridges (Fujita et al. 1974; Barnett 1981). In fusiform cambial cells, microtubules in the peripheral cytoplasm are oriented both longitudinally and horizontally with respect to the long axis of the cell (Srivastava 1966). Longitudinal microtubules are often observed close to the plasmalemma in fusiform cambial cells (Tsuda 1975; Rao 1985). The cortical microtubules are found in both active and dormant fusiform cambial cells but they are more abundant in active cambial cells as compared to dormant cambial cells (Itoh 1971; Tsuda

1975). Random arrays of cortical microtubules have been demonstrated in fusiform cambial cells by immuno-fluorescence microscopy (Abe et al. 1995a,b; Chaffey et al. 1997a,b,c; Funada et al. 1997). These observations were confirmed by TEM in ultrathin sections: cortical microtubules were arranged at various angles and occasionally overlapped (Chaffey et al. 1997a,d). The random arrays of cortical microtubules in fusiform cambial cells might be a general phenomenon in woody plants, regardless of whether they are conifers and hardwoods (Chaffey et al. 1997c).

A careful study of microtubule orientation in differentiating tracheids by immuno-fluorescence microscopy and confocal laser scanning microscopy has revealed that the predominant orientation of cortical microtubules is longitudinal with respect to the axis of the cell at the early stage of cell expansion (Fig. 3.5A; Abe et al. 1995a,b; Funada et al. 1997). The predominant orientation changes progressively from longitudinal to transverse during the radial expansion of cells. Finally, ordered and transversely oriented cortical microtubules are observed in tracheids at subsequent stages of differentiation, during which radial expansion ceases. Chaffey et al. (1997c) also found that the orientation of cortical microtubules changed from random within active fusiform cambial cells to helical within developing wood fibres. These observations indicate that the orientation and organization of cortical microtubules in differentiating tracheids and wood fibres changes successively during formation of the primary wall.

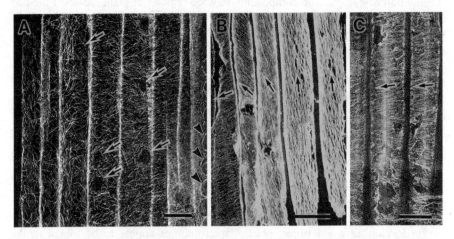

Fig. 3.5A-C. Immunofluorescence images, obtained by confocal laser scanning microscopy, showing the successive changes in the arrangement of cortical microtubules (viewed from the lumen side) in differentiating tracheids of *Abies sachalinensis*. **A** Cortical microtubules during formation of the primary wall. Cortical microtubules disappear locally (**arrows**) at sites of future intertracheal bordered pits, and circular bands of cortical microtubules (**arrowheads**) are visible around the edges of developing bordered pits. **B** and **C** Changes in the orientation of cortical microtubules (**arrows**) from a flat S-helix to a steep Z-helix to a flat S-helix during formation of the secondary wall. **Bar** 25 μm.

The progressive changes in the orientation of cortical microtubules resemble the reorientation of newly deposited cellulose microfibrils in tracheids during formation of the primary wall (Figs. 3.2 and 3.5A), suggesting a close relationship between the orientation of the cortical microtubules and the cellulose microfibrils. It appears that the cortical microtubules are involved in determining the orientation of newly deposited cellulose microfibrils by controlling the movement of the cellulose-synthase complexes in expanding tracheids. It has been proposed that cortical microtubules determine the direction of mechanical reinforcement by controlling the orientation of cellulose microfibrils (Green 1980). In elongating cells, cortical microtubules are arranged perpendicularly to the direction of elongation, whereas in multidirectionally expanding cells, cortical microtubules are oriented multidirectionally (Hogetsu and Oshima 1986; Hogetsu 1989; Kang et al. 1993). Thus, the orientation of cortical microtubules during the formation of primary walls reflects the direction of cell expansion. In the case of derivatives of fusiform cambial cells, lateral expansion might be the consequence of the predominantly longitudinal orientation of cellulose microfibrils in the primary wall. The orientation of these cellulose microfibrils in turn is controlled by the longitudinal cortical microtubules. Thus, the fusiform cambial cells are able to expand radially because cortical microtubules are arranged longitudinally. In contrast, in tracheids, the ordered and transversely oriented cellulose microfibrils on the innermost surface of the primary wall impede the lateral expansion of these cells. Their orientation might be controlled by cortical microtubules that are transverse as well.

Various stimuli, such as plant hormones, light, wounding, electrical fields and gravity, can shift the orientation of cortical microtubules in elongating cells (see Chap. 1). The effects of plant hormones on the reorientation of cortical microtubules have been particularly well documented (Shibaoka 1991, 1994). It has been demonstrated that auxin can induce the reorientation of cortical microtubules (Mayumi and Shibaoka 1996). Therefore, it seems likely that changes in the orientation of cortical microtubules from longitudinal to transverse during formation of the primary wall are mediated by endogenous plant hormones such as auxin. Recent studies in woody plants, by cryosectioning combined with microscale gas chromatography-mass spectrometry, have demonstrated a steep radial gradient in the level of indole-3-acetic acid (IAA), an endogenous auxin, across the cambial region that includes cambial cells, expanding cells and cells with thickening walls (Uggla et al. 1996; Tuominen et al. 1997). The level of endogenous IAA is maximal in the zone of cambial cells, i.e. in actively dividing cells. The level decreases through the zone of radially expanding xylem cells centripetally towards the zone of matured xylem cells. Thus, auxin appears to regulate cell division as well as the expansion of xylem cells, playing an important role in positional signaling. The level of IAA is low in cells that are in the process of secondary wall formation. This indicates the importance of factors other than auxin in control of secondary wall formation (Tuominen et al. 1997). It is possible that the orientation of cortical microtubules is determined by these radial gradients of auxin and possibly other plant hormones across the cambial region.

3.5
The interaction of microtubules and microfibrils in the secondary wall

The secondary wall in tracheids or wood fibres of conifers has a three-layered structure, with so-called S_1, S_2 and S_3 layers. In addition, detailed observations reveal that there are intermediate layers between the S_1 and the S_2 layer, and between the S_2 and the S_3 layers, respectively (Harada and Côté 1985). These layers are correlated to progressive changes in the angles of cellulose microfibrils relative to the cell axis. Thus, distinct boundaries between each of the layers cannot easily be distinguished.

When cell expansion ceases, well-ordered and transversely oriented cellulose microfibrils are observed on the innermost surfaces of the cell walls in differentiating tracheids of *Abies sachalinensis* (Abe et al. 1995a,b, 1997). These cellulose microfibrils are considered to belong to the secondary wall because of their texture. They are deposited just before birefringence becomes detectable by polarizing microscopy and cell expansion ceases (Abe et al. 1997). Thus, the radial expansion of tracheids might be restricted by the deposition of the secondary wall. A "crossed fibrillar texture" in the S_1 layer has been reported in tracheids (Wardrop 1957; Wardrop and Harada 1965; Harada and Côté 1985), and it was proposed that this structure might be due to alternations in S-helix and Z-helix in the S_1 layer. The helical direction, as observed from the outer face of the cells, is designated an S-helix (left-handed helix) or a Z-helix (right-hand helix) relative to the longitudinal axis of the cell. In contrast, Dunning (1968, 1969) proposed that the crossed fibrillar texture was merely due to a progressive change in the orientation of cellulose microfibrils within the S_1 layer. Recent observations indicate that the first-deposited, well-ordered cellulose microfibrils are oriented in a flat S-helix and then the orientation of newly deposited cellulose microfibrils changes from an S-helix to a flat Z-helix. (Kataoka et al. 1992; Prodhan et al. 1995a; Abe et al. 1997). This so-called S_1 layer exhibits a birefringence when transverse sections are observed by polarizing microscopy. The stage at which the S_1 layer forms is restricted to three or four differentiating wood fibres in a radial file during the season of active cambial growth (Prodhan et al. 1995a).

During formation of the secondary wall in tracheids or wood fibres, the cellulose microfibrils change their orientation progressively from a flat helix to a steep Z-helix, oriented at about 5-20° with respect to the cell axis, and with clockwise rotation when viewed from the lumen side (Fig. 3.6). This shift in the angles of cellulose microfibrils is considered to generate a semihelicoidal structure (Abe et al. 1991, 1995a; Prodhan et al. 1995a).

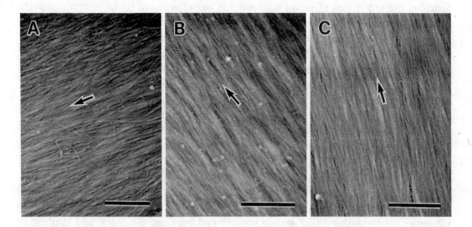

Fig. 3.6. Electronmicrographs (FE-SEM) of cellulose microfibrils (viewed from the lumen side) on the inner surface of the secondary wall during a similar stage in the differentiation of tracheids of *Abies sachalinensis* to that shown in Fig. 3.5B. The axes of tracheids in the photographs are vertical. The orientation of cellulose microfibrils (**arrows**) changes from a flat S-helix (**A**) over **B**, a flat Z-helix, to a steep Z-helix (**C**). **Bar** 0.5 μm.

The concept of a helicoidal pattern has been proposed for the cell walls of numerous plants (Roland and Vian 1979; Neville and Levy 1984; Roland et al. 1987; Vian and Reis 1991). The pattern consists of a series of planes in which the direction of cellulose microfibrils changes progressively. The arc-shaped or bow-shaped patterns observed by TEM in oblique ultrathin sections of tracheids or wood fibres correspond to the helicoidal structure (Parameswaran and Liese 1982; Roland and Mosiniak 1983; Prodhan et al. 1995b). Roland and Mosiniak (1983) demonstrated in wood fibres of *Tilia platyphyllos* that a rotational change in the orientation of cellulose microfibrils from the S_1 to the S_2 layer produced an intermediate twisted appearance, namely, a helicoidal texture.

Ten to fifteen tracheids or wood fibres with cellulose microfibrils in a steep Z-helix (Fig. 3.6C) are usually aligned in a radial file during the season of active cambial growth. No cellulose microfibrils with an S-helix are observed during formation of the S_2 layer. Meylan and Butterfield (1978) observed the direction of cellulose microfibrils in the S_2 layer in tracheids, wood fibres, and vessel elements of over 250 woody plants by SEM and concluded that the pattern was always a Z-helix. The cellulose microfibrils are closely aligned with the same orientation during the formation of the S_2 layer. The continuous deposition of cellulose microfibrils in one direction produces a thick cell-wall layer with a consistent texture. When the rotational change in the orientation of cellulose microfibrils is arrested, a thick cell-wall layer is formed as a result of the repeated deposition of cellulose microfibrils (Roland and Mosiniak 1983; Roland et al. 1987; Abe et al. 1991; Prodhan et al. 1995a). The thickness of the secondary wall is important in terms of the properties of wood because it is closely related to the specific gravity of wood, and specific gravity is positively correlated with mechanical strength (Panshin and de Zeeuw 1980; Zobel and van Buijitenen 1989). The duration of the arrest in the reorientation of cellulose microfibrils seems to determine the thickness of the S_2 layer and, thus, the thickness of the secondary wall: when the reorientation is arrested for a long period of time, the secondary wall becomes relatively thick. However, Kataoka et al. (1992) proposed another model, where the repetition of alternating changes in the direction of cellulose microfibrils results in a thick cell wall in tracheids. However, to date, a repeated alternation in the direction of cellulose microfibril reorientation has not been confirmed by other investigators.

The S_2 layer includes cellulose microfibrils in a steep Z-helix. The average microfibril angles are usually 5-30° with respect to the cell axis (Tsoumis 1991). However, these angles differ among species and between the radial and tangential wall, and depend also on the time of cell formation (earlywood and latewood; Saiki 1970; Harada and Côté 1985; Saiki et al. 1989). In addition, the microfibril angles in the S_2 layer can vary within the stem; the angles are usually large in the growth ring near the pith, namely, in the juvenile wood, and they decrease outwards to the bark side with increases in the age of cambium (Watanabe et al. 1963; Ohta et al. 1968; Donaldson 1992, 1993, 1996; Donaldson and Burdon 1995; Hirakawa and Fujisawa 1995; Hirakawa et al. 1997). Growth conditions may affect microfibril angles in the S_2 layer (Hiller and Brown 1967; Kubo 1983). Moreover, tracheids of compression wood that are formed on the lower side of inclined stems in conifers have large microfibril angles of about 45° with respect to the cell axis (Wardrop and Dadswell 1950; Timell 1986; Yoshizawa 1987). These variations in microfibril angles are due to differences in the reorientation of cellulose microfibrils. For example, while the direction of cellulose microfibrils in normal wood tracheids rearranges into a steep Z-helix that is oriented at about 5-20° with respect to the tracheid axis, the orientation of cellulose microfibrils in compression wood tracheids becomes oblique until the cellulose microfibrils are oriented in a Z-helix with an angle of about 45° with respect to the tracheid axis. Such a rotation of cellulose microfibrils might control the microfibril angle in the S_2 layer, and thus one of the most important factors that determines wood properties.

At the final stage of secondary-wall formation, the orientation of cellulose microfibrils changes from a steep Z-helix to a flat helix, with counterclockwise rotation when viewed from the lumen side. This corresponds to a directional switch in the orientation of the cellulose microfibrils from clockwise to counterclockwise (when viewed from the lumen side) during formation of the secondary wall. The deposition of cellulose microfibrils in a flat helix results in the S_3 layer. Within the S_3 layer, the cellulose microfibrils are deposited in bundles (Abe et al. 1991, 1994). The texture is different from that of the S_2 layer, where the cellulose microfibrils are neatly aligned. When transverse sections are observed by polarizing microscopy, the S_3 layer exhibits birefringence, as does the S_1 layer. The shift in angles of cellulose microfibrils is more abrupt during the transition from the S_2 to the S_3 layer as compared to the transition from the S_1 to the S_2 layer (Harada and Côté 1985; Abe et al. 1991). In general, the S_3 layer is thinner than the S_1 and S_2 layers. For instance, in earlywood, the thickness of radial walls of tracheids is 0.14-0.35 µm for the S_1, 0.98-1.93 µm for the S_2, and 0.07-0.15 µm for the S_3 layer; and in latewood it is 0.37-0.62 µm for the S_1, 2.13-6.94 µm for the S_2, and 0.08-0.14 µm for the S_3 layer (Saiki 1970). However, the absolute thickness differs between earlywood and latewood and among species.

The orientation of cellulose microfibrils on the innermost surface of the cell wall in tracheids or wood fibres varies widely depending on species (Liese 1960, 1965; Abe et al. 1992) and among individual tracheids (Abe et al. 1991). In addition, Wardrop (1964) reported that polarizing microscopy indicated that the S_3 layer was absent in some of the otherwise normal tracheids (e.g. *Picea*). Prodhan et al. (1995b) reported recently that cellulose microfibrils in juvenile wood were sparsely distributed on the innermost surfaces of wood fibres and, thus, no S_3 layer was formed that could be observed by polarizing microscopy. The change in orientation of cellulose microfibrils from the S_2 to the S_3 layer can also be considered to represent a helicoidal structure in the same way as the change from the S_1 to the S_2 layer (Roland and Mosiniak 1983; Roland et al. 1987; Abe et al. 1991). The difference in the rotation of cellulose microfibrils leads to variations in microfibrillar angles on the innermost surface of the cell wall in the tracheids or wood fibres. It is also well known that the S_3 layer is absent from the tracheids of compression wood in conifers (Timell 1986; Yoshizawa 1987) and from some gelatinous fibres of tension wood in hardwoods (the so-called $S_1 + S_2 +$ gelatinous layers and $S_1 +$ gelatinous layers types of hardwood; Dadswell and Wardrop 1955; Prodhan et al. 1995a,b). In these tracheids or wood fibres, the rotation of cellulose microfibrils might be limited to a single semihelicoidal pattern, as proposed by Roland et al. (1987).

As mentioned above, the direction of orientation of cellulose microfibrils changes progressively with changing speed of rotation during the formation of the secondary wall. Thus, if cortical microtubules control the orientation of cellulose microfibrils in the secondary wall in tracheids or wood fibres, cortical microtubules are expected to be parallel to the newly deposited cellulose microfibrils and thus to change their orientation progressively during formation of the secondary wall.

In differentiating xylem cells, cortical microtubules are abundant throughout the formation of the secondary wall (Barnett 1981), and numerous, parallel cortical microtubules, with a steep angle relative to the cell axis, are observed during formation of the secondary wall in wood fibres as well (Chaffey et al. 1997a). Cortical microtubules and cellulose microfibrils have been observed to be parallel in differentiating tracheids and wood fibres during the formation of the secondary wall in several woody plants (Cronshaw 1965; Nobuchi and Fujita 1972; Robards and Kidwai 1972; Fujita et al. 1974; Barnett 1981; Hirakawa 1984;

Inomata et al. 1992; Abe et al. 1994, 1995a; Prodhan et al. 1995a). Cronshaw (1965) observed a large number of cortical microtubules in developing wood fibres of *Acer rubrum*. When the S_2 layer is developing, parallel cortical microtubules are oriented in a helical direction and thus orientation is parallel to that of the cellulose microfibrils in the area with the most recent deposits of microfibrils. In oblique sections, cortical microtubules are seen to be oriented parallel to the orientation of cellulose microfibrils, as seen upon negative staining of the secondary wall in developing tracheids or wood fibres (Barnett 1981; Inomata et al. 1992). The cellulose microfibrils seem to be packed towards the inner face of the cell wall in almost the same direction as the cortical microtubules (Inomata et al. 1992).

During the formation of the secondary wall, following the cessation of cell expansion, the cortical microtubules are aligned in well-ordered arrays (Fig. 3.5b,c; Uehara and Hogetsu 1993; Abe et al. 1994, 1995a, b; Prodhan et al. 1995a; Chaffey et al. 1997a; Funada et al. 1997; Furusawa et al. 1998). Thus, two distinct arrangements of cortical microtubules are clearly detectable in differentiating tracheids or wood fibres: random arrays that are visible during the formation of primary walls and well-ordered arrays that are visible during the formation of secondary walls. The shift from random to well-ordered arrays seems to be a gradual process. During the successive steps of xylem differentiation, no tracheids or wood fibres with disassembled cortical microtubules can be seen in any radial files. The absence of disassembled microtubules indicates that the orientation of cortical microtubules might change progressively without complete depolymerization. Roberts et al. (1985) observed, in ethylene-treated pea epicotyl and mungbean hypocotyl, that the reorientation of cortical microtubules occurred from transverse to oblique to longitudinal without complete depolymerization. More recent observations, made after microinjection of neurotubulin with a fluorescent label into living plant cells, demonstrated that cortical microtubules reorient in a continuous and complex manner (Yuan et al. 1994, 1995; Lloyd et al. 1996). It has been proposed that the reorientation is initiated by the appearance of discordant cortical microtubules that do not share the existing alignment but anticipate the new direction. As the existing microtubules destabilize, the old alignment is gradually replaced by a new alignment. Although the process is not fully understood, modifications of microtubule-associated proteins might be related to the stability of cortical microtubules (see Chap. 1). The identification and cloning of these proteins are needed if we are to clarify the mechanisms related to the stability of cortical microtubules and, subsequently, the molecular basis of their reorientation.

Successive changes in the orientation of cortical microtubules can be observed in differentiating tracheids or wood fibres during the formation of secondary walls (Abe et al. 1994, 1995a; Prodhan et al. 1995a; Furusawa et al. 1998). The orientation changes by clockwise rotation from a flat S-helix to a steep Z-helix when viewed from the lumen side (Fig. 3.5B). This shift in the direction of cortical microtubules is completed within three or four tracheids or wood fibres in a radial file. Then, the cortical microtubules are oriented in a steep Z-helix at almost the same angle over the next 10 to 15 tracheids or wood fibres of the radial file. After further differentiation, the orientation of cortical microtubules returns from the steep Z-helix to a flat S-helix in tracheids or wood fibres (Fig. 3.5c). This shift is completed within one or two tracheids or wood fibres in a radial file.

These observations provide strong evidence for the hypothesis that the orientation of cortical microtubules changes progressively and in a manner similar to the changes in the orientation of newly deposited cellulose microfibrils during the formation of the secondary wall. Thus, there appears to be a very close relationship between cortical microtubules and newly deposited cellulose microfibrils. This is consistent with the idea that the cortical microtubules might control the ordered orientation of cellulose microfibrils in the semihelicoidal cell walls of tracheids or wood fibres in woody plants.

A similar parallelism between the orientation of cortical microtubules and cellulose microfibrils has been observed in the tension wood fibres of hardwoods. Tension wood fibres are usually characterized by the presence of gelatinous fibres, which form a gelatinous layer on the inner part of the cell wall. The gelatinous layer is composed entirely or almost entirely of cellulose microfibrils with a high degree of both crystallinity and parallelism, and the cellulose microfibrils are oriented parallel or nearly parallel to the cell axis. In gelatinous fibres of the S_1 + gelatinous layers type of hardwoods, such as *Fraxinus mandshurica* var. *japonica*, the cellulose microfibrils change their orientation from their original S-helix until they are oriented approximately parallel to the cell axis (Prodhan et al. 1995a). In parallel, the cortical microtubules in these fibres change their orientation from a flat helix to a direction parallel or nearly parallel to the cell axis during the formation of the secondary wall, similar to the rotation in the orientation of the cellulose microfibrils. Parallel or nearly parallel cortical microtubules are observed in differentiating gelatinous fibres of *Populus euroamericana* (Nobuchi and Fujita 1972; Fujita et al. 1974) and *Salix fragilis* (Robards and Kidwai 1972). During the formation of the gelatinous layer, the spacing of the cortical microtubules is close and their parallelism is pronounced. In contrast to normal tracheids or wood fibres, cortical microtubules do not change their orientation, in these gelatinous fibres, from parallel or nearly parallel to a flat helix at the final stage of secondary-wall formation. This absence of microtubule reorientation is mirrored on the level of cellulose microfibrils that remain parallel or nearly parallel to the cell axis in the inner layer of gelatinous fibres. These results support the proposed close relationship between cortical microtubules and cellulose microfibrils in the secondary wall of normal tracheids or wood fibres.

A parallel reorientation of cortical microtubules to that of the cellulose microfibrils has been observed in cotton fibres (Seagull 1992). During development, the cortical microtubules reorient from a random distribution to a shallow pitched helical pattern and then to a steeply pitched helical pattern. Coalignment of cortical microtubules and cellulose microfibrils can be observed throughout further development. However, slight differences between the orientation of cortical microtubules and cellulose microfibrils can be detected, leading Seagull (1992) to propose that the orientation of cellulose microfibrils, although under control of cortical microtubules, might be modified by interactions with components of the cell wall.

During the formation of the secondary wall, the orientation of cortical microtubules changes abruptly from a steep Z-helix to a flat S-helix, in contrast to the change from a flat S-helix to a steep Z-helix that is characteristic for tracheids or wood fibres. The shift in angles of newly deposited cellulose microfibrils is more abrupt during the transition from a Z-helix to a flat S-helix (i.e. from the S_2 to the S_3 layer) as compared to the transition from a flat S-helix to a steep Z-helix (i.e. from the S_1 to the S_2 layer). The velocity of microtubule reorientation might be related to the reorientation of newly deposited cellulose microfibrils. Therefore, the arrangements of cortical microtubules reflect the thickness of intermediate layers and the semihelicoidal pattern of the secondary wall.

Isolated single cells of *Zinnia* mesophyll that differentiate to tracheary elements can be used as experimental system to study wall formation in cell culture. In this system, an increase in the number of microtubules accompanies the reorganization of microtubules and the formation of the secondary wall (Fukuda 1994, 1996; Fukuda and Kobayashi 1989). In cotton fibres as well, the number of cortical microtubules increases fourfold early in the synthesis of the secondary wall as compared with the number during synthesis of the primary wall (Seagull 1992). This increase in the number of microtubules depends on the synthesis of tubulin de novo, that has been shown in the *Zinnia* system to be regulated at the transcriptional level (Fukuda 1994, 1996). In conifer tracheids, the density of cortical microtubules changes during formation of the secondary wall (Abe et al. 1994). In developing tracheids of *Abies sachalinensis* the average densities of cortical microtubules are 5.4 μm^{-1} of cell wall for the S_{12} layer (between the S_1 and the S_2 layer), 9.3 for the S_2 layer and 5.7 for the S_3 layer, respectively. This is reflected in the deposition of cellulose microfibrils that are more abundant during the formation of the S_2 layer stage than during the formation of the S_1 or the S_3 layers (Takabe et al. 1981a). The behaviour of cortical microtubules might be synchronized in relation to the active synthesis of cellulose. In developing gelatinous fibres during the formation of the gelatinous layer, the average densities of cortical microtubules range around 20 μm^{-1} (Fujita et al. 1974) and 17-18 μm^{-1} of cell wall (Prodhan et al. 1995a). These cortical microtubules are close to one another, with strong parallelism between them. The high density of cortical microtubules during formation of the S_2 and gelatinous layers might be related to the deposition of closely packed cellulose microfibrils in these layers. A recently proposed hypothesis suggests that cortical microtubules at high density are involved in the orientation of cellulose microfibrils while those at low density are not (Kimura and Mizuta 1994). These authors observed in green algae that cortical microtubules were oriented parallel to longitudinally deposited cellulose microfibrils correlated to a high average density of cortical microtubules of 216 microtubules per 50 μm of cell wall but they were never oriented parallel to transversely deposited cellulose microfibrils when the average density of cortical microtubules was 170 per 50 μm of cell wall. These results indicate that the distribution of microtubule density might regulate orientation and texture of cellulose microfibrils in the cell wall. Thus, the orientation of microtubules might be controlled by factors that regulate the synthesis of tubulin.

3.6
The influence of cortical microtubules on wood structure

Heterogeneous thickenings of the secondary wall are frequently observed in plant cells. These secondary thickenings are visible as annular, spiral, reticulate, scalari-form or pitted patterns. They are due to the localized deposition of components of the cell wall, in particular cellulose microfibrils. The localized deposition of cel-lulose microfibrils might be related to a heterogenous distribution of cellulose-synthase complexes (Herth 1985; Schneider and Herth 1986). Groups of cortical microtubules can be observed under ridges of secondary wall and are oriented parallel to the cellulose microfibrils and to the bands of the cell wall in developing tracheary elements (Hepler and Newcomb 1964; Pickett-Heaps 1974). Hardham and Gunning (1979) observed that groups of cortical microtubules appeared prior to the localized deposition of the secondary wall. Thus, the cortical microtubules might determine the pattern of the cell wall by defining the localization of cellu-lose microfibrils in the cell wall. By contrast, Hogetsu (1991) proposed that the cortical microtubules might determine and maintain a boundary between plasma membrane domains where cellulose-synthase complexes are activated and do-mains where they are inactive, thereby defining the sites of localized deposition of cellulose microfibrils.

The cells of the secondary xylem of woody plants include modifications in structure that are normal features of the cell wall (Panshin and de Zeeuw 1980; Ohtani 1994). These modifications, such as pits, perforations, spiral thickenings and warts, are formed by localized deposition of cell wall materials. Their size, structure and number are characteristic anatomical features for individual species and, thus, they are frequently used for the identification of woods.

Gaps in the secondary wall of secondary xylem cells, the so called pits, are im-portant for the cell-to-cell movement of water, nutrients, micro- and macromole-cules and ions. Tracheids form bordered pits that are characterized by a membrane that is overarched by the secondary wall. Concentrical rings of cellulose microfi-brils on the inner surface of pit borders form the border thickening (Liese 1965; Harada and Côté 1985). In differentiating tracheids at the final stage of cell expan-sion, circular bands of cortical microtubules are observed around the edges of developing bordered pits (Figs. 3.5A and 3.7a; Hirakawa and Ishida 1981; Uehara and Hogetsu 1993; Abe et al. 1995b; Funada et al. 1997). Similar circular bands of cortical microtubules around the pit aperture of bordered pits can be seen in differentiating vessel elements (Robards and Humpherson 1967; Chaffey et al. 1997b). In these regions, the pit borders are clearly visible by differential interference contrast (Uehara and Hogetsu 1993; Funada et al. 1997). As tracheids differentiate, the circular bands of cortical microtubules narrow centripetally and finally they disappear upon completed development of the bordered pits. Thus, the circular bands of cortical microtubules seem to be involved in the deposition of concentrically oriented cellulose microfibrils at pit borders. In addition, the circular bands of cortical microtubules around pits might determine and maintain the boundary between the forming pit and the neighbouring areas of the plasma membrane (Hogetsu 1991; Uehara and Hogetsu 1993).

Prior to the appearance of circular bands of cortical microtubules, during early stages in the formation of the primary wall, the cortical microtubules in differentiating tracheids disappear locally from areas of prospective intertracheal bordered pits (Fig. 3.5A; Abe et al. 1995b; Funada et al. 1997). Enlarged perforations within the reticulum of cortical microtubules are also detected at the sites of the prospective pits in differentiating vessel elements of *Aesculus hippocastanum* (Chaffey et al. 1997b). Thus, the localized disappearance of cortical microtubules might be involved in the commitment of a given area to pit formation. The pit-forming region might be delineated from non-pit regions by localized cortical microtubules during the early stages of primary-wall formation (Funada et al. 1997). These results are consistent with the hypothesis that, in tracheids, the commitment for pit formation begins much earlier than the formation of the secondary wall (Barnett and Harris 1975; Barnett 1981), and that the bordered pits are, in fact, delineated already in the primary wall (Bauch et al. 1968; Fengel 1972; Imamura and Harada 1973). The involvement of cortical microtubules in the determination and maintenance of the positions of pit borders emphasizes again the importance of cortical microtubules for the localized deposition of cell wall materials.

The innermost surface of the secondary wall of xylem cells develops localized ridges made up of parallel bundles of cellulose microfibrils. These cellulose microfibrils are oriented helically with respect to the cell axis. Such cell-wall thickenings are known as spiral thickenings. Spiral thickenings are observed in tracheids or wood fibres of only a few of species but they are relatively common in vessel elements. During final stages of secondary-wall formation, obliquely oriented bands of cortical microtubules are observed in tracheids of conifers, in which spiral thickenings are formed (Uehara and Hogetsu 1993; Furusawa et al. 1998). These bands of cortical microtubules are approximately 3-4 μm in width and then condense into rope-like structures. These rope-like bundles of cortical microtubules are arranged helically underneath the cell wall around the tracheid (Fig. 3.7b). At the time when these rope-like bundles of cortical microtubules are observed, helical thickenings become visible by differential interference contrast (Fig. 3.7c). The rope-like bundles of cortical microtubules can be superimposed on these helical wall thickenings.

Fig. 3.7A-C. Immunofluorescence images (**A,B**) and Nomarski differential interference contrast image (**c**), obtained by confocal laser scanning microscopy showing arrangements of cortical microtubules that occur during the modification of wood structure in *Taxus cuspidata*. **A** Circular bands of cortical microtubules (**arrow**) are visible around the edges of developing bordered pits. **B,C** Bands of helically oriented cortical microtubules (**arrows** in **b**) are superimposed on helical thickenings (**arrows** in **C**). **Bar** 25 μm.

This spatial congruence suggests that cortical microtubules might be involved in the formation of helical thickenings and thus, again, they might control the localized deposition of cellulose microfibrils. In the final stage of differentiation, different arrays of cortical microtubules can be observed within the same tracheid (Furusawa et al. 1998). For instance, microtubules can be localized in a band-like pattern in one part of such tracheids, with disordered cortical microtubules between the bands, whereas in another part of the tracheid, the helical bundles of cortical microtubules can be observed with few cortical microtubules being interspersed between the bands. These tracheids might represent transitional states during the dynamic reorganization of cortical microtubules. The coexistence of different arrays suggests that changes in the arrangement of cortical microtubules can occur non-uniformly within a single tracheid.

The results described above indicate that the localized appearance or disappearance of cortical microtubules controls the localized deposition of cellulose microfibrils, resulting in modifications of the cell wall. The mechanisms for the reorientation and localization of cortical microtubules are not fully understood. Kobayashi et al. (1987, 1988) and Fukuda and Kobayashi (1989) proposed that actin filaments might play an important role in shifting the orientation and localization of microtubules in differentiating tracheary elements of cultured *Zinnia* cells. They observed that reticulate bundles of microtubules and aggregates of actin filaments between microtubules emerged simultaneously from sites that are adjacent to the plasma membrane just before the formation of the secondary wall. Subsequently, aggregates of actin filaments extended transversely to the cell axis and microtubules became transversely aligned between actin filaments as well.

Transverse ridges of secondary wall are formed above the transverse bundles of microtubules in these tracheary elements (Falconer and Seagull 1985). The disruption of actin filaments by cytochalasin B prevents the reorientation of microtubules into transverse arrays. These results suggest that actin filaments might regulate the orientation and localization of microtubules. The presence of bundles of microfilaments (actin filaments) has been reported in cambial cells of woody plants (Srivastava 1966; Srivastava and O'Brien 1966; Tsuda 1975; Goosen-de Roo et al. 1983; Chaffey et al. 1997a). The microfilaments are generally oriented longitudinally. The lengths of microfilament bundles range from 7 to 17 µm (Catesson 1990) and the mean number of microfilaments per bundle is 16 (Goosen-de Roo et al. 1983). Chaffey et al. (1997a) recently observed in *Aesculus hippocastanum* that bundles of axially oriented microfilaments were present in fusiform cambial cells and, moreover, the longitudinal orientation of microfilaments was retained in cambial derivatives during xylem differentiation even though the orientation of cortical microtubules had changed. These observations suggest that microfilaments might not have a major role in the change in orientation of cortical microtubules in differentiating cells derived from cambial cells. Chaffey et al. (1997a) pointed out that the behaviour of cortical microtubules and microfilaments in a natural system such as the vascular differentiation of the cambium might differ from in vitro systems, such as the *Zinnia* system. Much work remains to be done to elucidate the role of actin filaments in controlling the orientation and localization of cortical microtubules in the natural system, i.e. xylem differentiation.

3.7
Potential approaches to the control of microfibril orientation

Most properties of wood are moderately to strongly inherited (Zobel and van Buijtenen 1989; Zobel and Jett 1995). This high heritability shows that improvements in wood quality by selective breeding should be more effective than a silvicultural approach. Thus, it is feasible to design breeding programs that might allow genetical control of wood properties. However, despite intensive breeding efforts this approach has not led to improved wood quality, since most programs for tree improvement have focused on growth rate, stem form and resistance to diseases and insects. Zobel and Jett (1995) emphasized the future importance of genetic manipulation in efforts to improve the quality and uniformity of wood. Recent observations have shown that microfibril angles in the secondary wall might be genetically controlled (Donaldson and Burdon 1995; Hirakawa and Fujisawa 1995). Variations in microfibril angles from the pith to the bark in *Cryptomeria japonica* exhibited almost the same pattern in all clones examined: angles were large in the growth rings close to the pith and became almost constant at a distance of about 15 growth rings from the pith (Hirakawa and Fujisawa 1995). However, the microfibril angles differed significantly among clones of different origin (Donaldson and Burdon 1995; Hirakawa and Fujisawa 1995). Since a clone is derived from a single original tree by vegetative propagation, all cloned trees of a single origin are genetically identical. Thus, differences in wood properties among clones of different origin must be due to differences in genetic factors. Donaldson and Burdon (1995) reported the relatively high value of 0.7 for the clonal repeatability (an estimate of broad heritability) of microfibril angles in *Pinus radiata*. By contrast, both site and growth rate had little effect on microfibril angles (Donaldson 1996). These observations indicate that the manipulation of microfibril angles by selective breeding has a higher potential for improvement of wood than attempts that are based on silvicultural treatments. Thus, detailed screens to select plus trees with low microfibril angles (i.e. longitudinal microfibrils) should be included in future breeding programs. In addition, the development of efficient methods for vegetative propagation is needed for the genetic improvement of wood quality. Many woody plants, in particular some conifers, are still resilient not only to traditional vegetative propagation, such as cuttings, but also to micropropagation (Bonga and von Aderkas 1992; Zobel and Jett 1995).

As mentioned above, a close relationship exists between the orientation and localization of cortical microtubules and the orientation and localization of newly deposited cellulose microfibrils in differentiating tracheids or wood fibres, indicating that cortical microtubules control the organization of cellulose microfibrils in the cell wall. In addition, microfibril angles in the secondary wall might also be genetically controlled. Thus, it is possible that the behaviour of cortical microtubules, such as the angle to the cell axis and the rotational speed during formation of the secondary wall, might also be controlled by genetic factors. In plus trees with low microfibril angles (i.e. longitudinal microfibrils), cortical microtubules probably change their orientation until they are oriented at a very steep angle to the cell axis during formation of the S_2 layer. Thus, the genetic manipulation of cortical microtubules during formation of the S_2 layer might help to control microfibril angles of tracheids or wood fibres. Furthermore, detailed comparisons among clones with different microfibril angles might help to understand the genetic mechanism of organization of cortical microtubules during formation of the secondary wall.

The development of techniques for genetic transformation and regeneration of trees is also important for the manipulation of wood quality (Cheliak and Rogers 1990; Jouanin et al. 1993). Recently, transgenic hybrid aspen (*Populus tremula* x *Populus tremuloides*) trees expressing the IAA-biosynthetic genes *iaaA* and *iaaH* from *Agrobacterium tumefaciens* and the *rolC* gene from *Agrobacterium rhizogenes* were produced (Tuominen et al. 1995, 1997; Nilsson et al. 1996). These trees exhibited changes not only in growth pattern, including tree height, stem diameter, and stem fasciation, but also in the anatomical features of xylem cells, such as vessel size, vessel density and lumen area of wood fibres, together with changed balance and metabolism of endogenous plant hormones. Although these authors did not describe the cell wall structures in these transgenic trees in detail, the possibility exists that the ultrastructure of the cell wall of xylem cells, including microfibril angles, might also have been altered, since it is well known that plant hormones reorient cortical microtubules (Shibaoka 1991,1994). Modifications in the levels and distribution of endogenous plant hormones by genetic engineering might be a very powerful strategy for modifying wood quality. In addition, such transgenic trees would be excellent models for studies of the physiological mechanism of wood formation.

There is considerable evidence that various stimuli other than plant hormones can also induce the reorientation of cortical microtubules in elongating plant cells (Shibaoka 1991, 1994). In artificially inclined trees, gravity changes the pattern of alignment of cortical microtubules in differentiating tracheids or wood fibres after the cessation of cell expansion (Prodhan et al. 1995a; Furusawa et al. 1998). Thus, external stimuli can alter the orientation of cortical microtubules during formation of the secondary wall as well as the primary wall in elongating cells. The reorientation of cortical microtubules induced by an external stimulus must involve several steps, namely, perception of the stimulus, signal transduction, a change in directional factors and reorientation of cortical microtubules. The molecular mechanisms of these processes are largely unknown (see Chap. 1). The micromanipulation of directional factors that control the direction and extent of change in the orientation of cortical microtubules might be specific for the modification of microfibril angles. For this approach, both the molecular characterization of the various factors and the characterization of their specific regulation of features of the secondary wall are necessary. Biotechnological control of microfibrillar angles and of the thickness of the secondary wall in tracheids or wood fibres by manipulation of cortical microtubules could provide new tools that would permit the control of wood quality.

Acknowledgements. The author thanks Dr. J. Ohtani, Dr. K. Fukazawa, Dr. H. Abe, Dr. A.K.M.A. Prodhan, Mr. O. Furusawa, Mr. H. Miura, Mr. M. Shibagaki and Miss H. Imaizumi for their cooperation during the preparation of this chapter. The author's research has been supported by Grants-in-Aid for Scientific Research from the Ministry of Education, Science and Culture, Japan (Nos. 06404013 and 09760152) and the Japan Society for the Promotion of Science (No. JSPS-RFTF 96L00605).

References

Abe H, Ohtani J, Fukazawa K (1991) FE-SEM observation on the microfibrillar orientation in the secondary wall of tracheids. IAWA Bull 12: 431-438
Abe H, Ohtani J, Fukazawa K (1992) Microfibrillar orientation of the innermost surface of conifer tracheid walls. IAWA Bull 13: 411-417
Abe H, Ohtani J, Fukazawa K (1994) A scanning electron microscopic study of changes in microtubule distributions during secondary wall formation in tracheids. IAWA Bull 15: 185-189
Abe H, Funada R, Imaizumi H, Ohtani J, Fukazawa K (1995a) Dynamic changes in the arrangement of cortical microtubules in conifer tracheids during differentiation. Planta 197: 418-421
Abe H, Funada R, Ohtani J, Fukazawa K (1995b) Changes in the arrangement of microtubules and microfibrils in differentiating conifer tracheids during the expansion of cells. Ann Bot 75: 305-310
Abe H, Funada R, Ohtani J, Fukazawa K (1997) Changes in the arrangement of cellulose microfibrils associated with the cessation of cell expansion in tracheids. Trees 11: 328-332

Aloni R (1987) Differentiation of vascular tissues. Annu Rev Plant Physiol 38: 179-204

Baïer M, Goldberg R, Catesson AM, Liberman M, Bouchemal N, Michon V, Hervé du Penhoat C (1994) Pectin changes in samples containing poplar cambium and inner bark in relation to the seasonal cycle. Planta 193: 446-454

Bailey IW (1920) The cambium and its derivative tissue 2. Size variations of cambial initials in gymnosperms and angiosperms. Am J Bot 7: 355-367

Bailey IW, Vestal MR (1937) The orientation of cellulose in the secondary wall of tracheary cells. J Arnold Arbor 18: 185-195

Barber NF, Meylan BA (1964) The anisotropic shrinkage of wood. Holzforschung 18: 146-156

Barnett JR (1981) Secondary xylem cell development. In: Barnett JR (ed) Xylem cell development. Castle House Publications, Tunbridge Wells, pp 47-95

Barnett JR, Harris JM (1975) Early stages of bordered pit formation in radiata pine. Wood Sci Technol 9: 233-241

Bauch J, Liese W, Scholz JM (1968) Über die Entwicklung und stoffliche Zusammensetzung der Hoftüpfelmembranen von Längstracheiden in Coniferen. Holzforschung 22: 144-153

Bonga JM, von Aderkas P (1992) In vitro culture of trees. Kluwer, Dordrecht, pp 1-236

Catesson AM (1974) Cambial cells. In: Robards AW (ed) Dynamic aspects of plant ultrastructure. MacGraw-Hill, London, pp 358-390

Catesson AM (1990) Cambial cytology and biochemistry. In: Iqbal M (ed) The vascular cambium. Research Studies Press, Taunton, pp 63-112

Catesson AM (1994) Cambial ultrastructure and biochemistry: changes in relation to vascular tissue differentiation and the seasonal cycle. Int J Plant Sci 155: 251-261

Catesson AM, Funada R, Robert-Baby D, Quinet-Szèly M, Chu-Bâ J, Goldberg R (1994) Biochemical and cytochemical cell wall changes across the cambial zone. IAWA J 15: 91-101

Cave ID (1966) Theory of X-ray measurement of microfibril angle in wood. For Prod J 16: 37-42

Cave ID (1968) The anisotropic elasticity of the plant cell wall. Wood Sci Technol 2: 268-278

Cave ID, Walker JCF (1994) Stiffness of wood in fast-grown plantation softwoods: the influence of microfibril angle. For Prod J 44: 43-48

Chaffey NJ, Barlow P, Barnett J (1997a) Cortical microtubules rearrange during differentiation of vascular cambial derivatives, microfilaments do not. Trees 11: 333-341

Chaffey NJ, Barnett JR, Barlow PW (1997b) Cortical microtubule involvement in bordered pit formation in secondary xylem vessel elements of *Aesculus hippocastanum* L. (Hippocastanaceae): a correlative study using electron microscopy and indirect immunofluorescence microscopy. Protoplasma 197: 64-75

Chaffey NJ, Barnett JR, Barlow PW (1997c) Visualization of the cytoskeleton within the secondary vascular system of hardwood species. J Microsc 187: 77-84

Chaffey N, Barnett J, Barlow P (1997d) Endomembranes, cytoskeleton, and cell wall: aspects of the ultrastructure of the vascular cambium of taproots of *Aesculus hippocastanum* L. (Hippocastanaceae). Int J Plant Sci 158: 97-109

Cheliak WM, Rogers DL (1990) Integrating biotechnology into tree improvement programs. Can J For Res 20: 452-463

Côté WA, Hanna RB (1985) Trends in application of electron microscopy to wood research. In: Kucera LJ (ed) Xylorama. Birkhäuser, Basel, pp 42-50

Cronshaw J (1965) Cytoplasmic fine structure and cell wall development in differentiating xylem elements. In: Côté WA jr (ed) Cellular ultrastructure of woody plants. Syracuse University Press, Syracuse, pp 99-124

Dadswell HE, Wardrop AB (1955) The structure and properties of tension wood. Holzforschung 9: 97-104

Donaldson LA (1992) Within- and between-tree variation in microfibril angle in *Pinus radiata*. N Z J For Sci 22: 77-86

Donaldson LA (1993) Variation in microfibril angle among three genetic groups of *Pinus radiata* trees. N Z J For Sci 23: 90-100

Donaldson LA (1996) Effect of physiological age and site on microfibril angle in *Pinus radiata*. IAWA J 17: 421-429

Donaldson LA, Burdon RD (1995) Clonal variation and repeatability of microfibril angle in *Pinus radiata*. New Zeal J For Sci 25: 164-174

Dunning CE (1968) Cell-wall morphology of longleaf pine latewood. Wood Sci 1: 65-76

Dunning CE (1969) The structure of longleaf pine latewood 1. Cell wall morphology and the effect of alkaline extraction. Tappi 52: 1326-1335

Falconer MM, Seagull RW (1985) Immunofluorescent and calcofluor white staining of developing tracheary elements in *Zinnia elegans* L. suspension cultures. Protoplasma 125: 190-198

Fengel D (1972) Structure and function of the membrane in softwood bordered pits. Holzforschung 26: 1-9

Fujii T, Harada H, Saiki H (1977) Ultrastructure of expanding ray parenchyma cell wall in poplar (*Populus koreana* Rehd.). Bull Kyoto Univ For 49: 127-131

Fujita M, Saiki H, Harada H (1974) Electron microscopy of microtubules and cellulose microfibrils in secondary wall formation of poplar tension wood fibres. Mokuzai Gakkaishi 20: 147-156

Fujita M, Saiki H, Harada H (1978) The secondary wall formation of compression wood tracheids 3: Cell organelles in relation to cell wall thickening and lignification. Mokuzai Gakkaishi 24: 353-361

Fukuda H (1994) Redifferentiation of single mesophyll cells into tracheary elements. Int J Plant Sci 155: 262-271

Fukuda H (1996) Xylogenesis: initiation, progression, and cell death. Annu Rev Plant Physiol Plant Mol Biol 47: 299-325

Fukuda H, Kobayashi H (1989) Dynamic organization of the cytoskeleton during tracheary-element differentiation. Dev Growth Differ 31: 9-16

Funada R, Catesson AM (1991) Partial cell wall lysis and the resumption of meristematic activity in *Fraxinus excelsior* cambium. IAWA Bull 12: 439-444

Funada R, Abe H, Furusawa O, Imaizumi H, Fukazawa K, Ohtani J (1997) The orientation and local-ization of cortical microtubules in differentiating conifer tracheids during cell expansion. Plant Cell Physiol 38: 210-212

Furusawa O, Funada R, Murakami Y, Ohtani J (1998) The arrangement of cortical microtubules in compression wood tracheids of *Taxus cuspidata* visualized by confocal laser microscopy. J Wood Sci 44: 230-233

Giddings TH Jr, Staehelin LA (1988) Spatial relationship between microtubules and plasma-membrane rosettes during the deposition of primary wall microfibrils in *Closterium* sp. Planta 173: 22-30

Giddings TH Jr, Staehelin LA (1991) Microtubule-mediated control of microfibril deposition: a re-examination of the hypothesis. In: Lloyd CW (ed) The cytoskeletal basis of plant growth and form. Academic Press, London, pp 85-99

Goosen-de Roo L, van Spronsen PC (1978) Electron microscopy of the active cambial zone of *Frax-inus excelsior*. IAWA Bull ns 1978: 59-64

Goosen-de Roo L, Burggraaf PD, Libbenga KR (1983) Microfilament bundles associated with tubular endoplasmic reticulum in fusiform cells in the active cambial zone of *Fraxinus excelsior* L. Proto-plasma 116: 204-208

Green PB (1980) Organogenesis – a biophysical view. Annu Rev Plant Physiol 31: 51-82

Green PB, King A (1966) A mechanism for the origin of specifically oriented textures with special reference to *Nitella* wall texture. Aust J Biol Sci 19: 421-437

Guglielmino N, Liberman M, Catesson AM, Mareck A, Prat R, Mutaftschiev S, Goldberg R (1997) Pectin methylesterases from poplar cambium and inner bark: localization properties and seasonal changes. Planta 202: 70-75

Gunning BES, Hardham RH (1982) Microtubules. Annu Rev Plant Physiol 33: 651-698

Harada H (1965) Ultrastructure and organization of gymnosperm cell walls. In: Côté WA jr (ed) Cel-lular ultrastructure of woody plants. Syracuse University Press New York, pp 215-233

Harada H, Côté WA jr (1985) Structure of wood. In: Higuchi T (ed) Biosynthesis and biodegradation of wood components. Academic Press Orlando, pp 1-42

Hardham AR, Gunning BES (1978) Interpolation of microtubules into cortical arrays during cell elongation and differentiation in roots of *Azolla pinnata*. J Cell Sci 37: 411-442

Harris JM, Meylan BA (1965) The influence of microfibril angle on longitudinal and tangential shrinkage in *Pinus radiata*. Holzforschung 19: 144-153

Heath IB (1974) A unified hypothesis for the role of membrane bound enzyme complexes and microtubules in plant cell wall synthesis. J Theor Biol 48: 445-449

Heath IB, Seagull RW (1982) Oriented cellulose fibrils and the cytoskeleton: a critical comparison of models. In: Lloyd CW (ed) The cytoskeleton in plant growth and development. Academic Press, London, pp 163-182

Hepler PK, Newcomb EH (1964) Microtubules and fibrils in the cytoplasm of *Coleus* cells undergoing secondary wall deposition. J Cell Biol 20: 529-533

Hepler KH, Palevitz BA (1974) Microtubules and microfilaments. Annu Rev Plant Physiol 25: 309-362

Herth W (1985) Plasma membrane rosettes involved in localized wall thickening during xylem vessel formation of *Lepidium sativum* L. Planta 164: 12-21

Higuchi T (1997) Biochemistry and molecular biology of wood. Springer, Berlin Heidelberg New York, pp 1-362

Hiller CH, Brown RS (1967) Comparison of dimensions and fibril angles of loblolly pine tracheids formed in wet or dry growing seasons. Am J Bot 54: 453-460

Hirakawa Y (1984) A SEM observation of microtubules in xylem cells forming secondary walls of trees. Res Bull College Exp For Hokkaido Univ 41: 535-550

Hirakawa Y, Fujisawa Y (1995) The relationship between microfibril angles of the S_2 layer and latewood tracheid lengths in elite sugi tree (*Cryptomeria japonica*) clones. Mokuzai Gakkaishi 41: 123-131

Hirakawa Y, Ishida S (1981) A scanning and transmission electron microscopic study of layered structure of wall in pit border region between earlywood tracheids in conifer. Res Bull College Exp For Hokkaido Univ 38: 249-264

Hirakawa Y, Yamashita K, Nakada R, Fujisawa Y (1997) The effects of S_2 microfibril angles of latewood tracheids and densities on modulus of elasticity variations of sugi tree (*Cryptomeria japonica*) logs. Mokuzai Gakkaishi 43: 717-724

Hogetsu T (1989) The arrangement of microtubules in leaves of monocotyledonous and dicotyledonous plants. Can J Bot 67: 3506-3512

Hogetsu T (1991) Mechanism for formation of secondary wall thickening in tracheary elements: microtubules and microfibrils of tracheary elements of *Pisum sativum* L. and *Commelina communis* L. and the effects of amiprophosmethyl. Planta 185: 190-200

Hogetsu T, Oshima Y (1986) Immunofluorescence microscopy of microtubule arrangement in root cells of *Pisum sativum* L. var. Alaska. Plant Cell Physiol 27: 939-945

Hogetsu T, Shibaoka H (1978) Effects of colchicine on cell shape and on microfibril arrangement in the cell wall of *Closterium acerosum*. Planta 140: 15-18

IAWA Committee (1964) Multilingual glossary of terms used in wood anatomy. Konkordia, Winterthur, pp 1-186

Imagawa H (1981) Study on the seasonal development of the secondary phloem in *Larix leptolepis*. Res Bull College Exp For Hokkaido Univ 38: 31-44

Imamura Y, Harada H (1973) Electron microscopic study on the development of the bordered pit in coniferous tracheids. Wood Sci Technol 7: 189-205

Imamura Y, Harada H, Saiki H (1972) Electron microscopic study on the formation and organization of the cell wall in coniferous tracheids. Bull Kyoto Univ For 44: 183-193

Inomata F, Takabe K, Saiki H (1992) Cell wall formation of conifer tracheid as revealed by rapid-freeze and substitution method. J Electron Microsc 41: 369-374

Itoh T (1971) On the ultrastructure of dormant and active cambium of conifers. Wood Res 51: 33-45

Itoh T (1991) The formation of plant cell wall – structural approach. Mokuzai Gakkaishi 37: 775-789

Jouanin L, Brasileiro ACM, Leplé JC, Pilate, G, Cornu D (1993) Genetic transformation: a short review of methods and their applications, results and perspectives for forest trees. Ann Sci For 50: 325-336

Kang KD, Itoh T, Soh WY (1993) Arrangement of cortical microtubules in elongating epicotyl of *Aesculus turbinata* Blume. Holzforschung 47: 9-18

Kataoka Y, Saiki H, Fujita M (1992) Arrangement and superimposition of cellulose microfibrils in the secondary walls of coniferous tracheids. Mokuzai Gakkaishi 38: 327-335

Kerr T, Bailey IW (1934) The cambium and its derivative tissues. 5. Structure, optical properties and chemical composition of the so-called middle lamella. J Arnold Arbor 15: 327-349

Kidwai P, Robards AW (1969) The appearance of differentiating vascular cells after fixation in different solutions. J Exp Bot 20: 664-670

Kimura S, Mizuta S (1994) Role of the microtubule cytoskeleton in alternating changes in cellulose microfibril orientation in the coenocytic green alga *Chaetomorpha moniligera*. Planta 193: 21-31

Kobayashi H, Fukuda H, Shibaoka H (1987) Reorganization of actin filaments associated with the differentiation of tracheary elements in *Zinnia* mesophyll cells. Protoplasma 138: 69-71

Kobayashi H, Fukuda H, Shibaoka H (1988) Interrelation between the spatial disposition of actin filaments and microtubules during the differentiation of tracheary elements in cultured *Zinnia* cells. Protoplasma 143: 29-37

Kubo T (1983) Annual ring structure and its formation in sugi (*Cryptomeria japonica*) having different amounts of crown. 6. Structural characteristics. Mokuzai Gakkaishi 29: 725-730

Kutschera U (1991) Regulation of cell expansion. In: Lloyd CW (ed) The cytoskeletal basis of plant growth and form. Academic Press, London, pp 149-158

Lang JM, Eisinger WR, Green PB (1982) Effects of ethylene on the orientation of microtubules and cellulose microfibrils of pea epicotyls with poly-lamellated walls. Protoplasma 110: 5-14

Larson PR (1994) The Vascular cambium: development and structure. Springer, Berlin Heidelberg New York, pp 1-725

Ledbetter MC, Porter KR (1963) A "microtubule" in plant cell fine structure. J Cell Biol 19: 239-250

Liese W (1960) Die Struktur der Tertiärwand in Tracheiden und Holzfasern. Holz Roh- Werkstoff 18: 296-303

Liese W (1965) The fine structure of bordered pits in softwoods. In: Côté WA Jr (ed) Cellular ultrastructure of woody plants. Syracuse University Press, New York, pp 271-290

Lloyd CW (1987) The plant cytoskeleton: the impact of fluorescence microscopy. Annu Rev Plant Physiol 38: 119-139

Lloyd CW, Slabas AR, Powell AJ, Macdonald G, Badley RA (1979) Cytoplasmic microtubules of higher plant cells visualised with anti-tubulin antibodies. Nature 279: 239-241

Lloyd CW, Shaw PJ, Warn RM, Yuan M (1996) Gibberellic-acid-induced reorientation of cortical microtubules in living plant cells. J Microsc 181: 140-144

Mayumi K, Shibaoka H (1996) The cyclic reorientation of cortical microtubules on walls with a crossed polylamellate structure: effects of plant hormones and an inhibitor of protein kinases on the progression of the cycle. Protoplasma 195: 112-122

Meylan BA, Butterfield BG (1978) Helical orientation of the microfibrils in tracheids fibres and vessels. Wood Sci Technol 12: 219-222

Neville AC, Levy S (1984) Helicoidal orientation of cellulose microfibrils in *Nitella opaca* internode cells: ultrastructure and computed theoretical effects of strain reorientation during wall growth. Planta 162: 370-384

Nilsson O, Moritz T, Sundberg B, Sandberg G, Olsson O (1996) Expression of the *Agrobacterium rhizogenes rolC* gene in a deciduous forest tree alters growth and development and leads to stem fasciation. Plant Physiol 112: 493-502

Nobuchi T, Fujita M (1972) Cytological structure of differentiating tension wood fibres of *Populus euroamericana*. Mokuzai Gakkaishi 18: 137-144

Ohta S, Watanabe H, Matsumoto T, Tsutsumi J (1968) Studies on mechanical properties of juvenile wood. 2. Variation of fundamental structural factors and mechanical properties of hinoki trees (*Chamaecyparis obtusa* Sieb. et Zucc.). Mokuzai Gakkaishi 14: 261-268

Ohtani J (1994) Modifications of the wood cell wall. Mokuzai Gakkaishi 40: 1275-1283

Panshin AJ, de Zeeuw C (1980) Textbook of wood technology, 4th edn. MacGraw-Hill, New York, pp 1-722

Parameswaran N, Liese W (1982) Ultrastructural localization of wall components in wood cells. Holz Roh- Werkstoff 40: 145-155

Pickett-Heaps JD (1974) Plant microtubules. In: Robards AW (ed) Dynamic aspects of plant ultra-structure. MacGraw-Hill, London, pp 219-255

Prodhan AKMA, Funada R, Ohtani J, Abe H, Fukazawa K (1995a) Orientation of microfibrils and microtubules in developing tension-wood fibres of Japanese ash (*Fraxinus mandshurica* var. *japonica*). Planta 196: 577-585

Prodhan AKMA, Ohtani J, Funada R, Abe H, Fukazawa K (1995b) Ultrastructural investigation of tension wood fibre in *Fraxinus mandshurica* Rupr. Var. *japonica* Maxim. Ann Bot 75: 311-317

Rao KS (1985) Seasonal ultrastructural changes in the cambium of *Aesculus hippocastanum* L. Ann Sci Nat Bot Veg Ser 137: 213-228

Ridge I (1973) The control of cell shape and rate of cell expansion by ethylene: effects on microfibril orientation and cell extensibility in etiolated peas. Acta Bot Neerl 22: 144-158

Robards AW, Humpherson PG (1967) Microtubules and angiosperm bordered pit formation. Planta 77: 233-238

Robards AW, Kidwai PA (1972) Microtubules and microfibrils in xylem fibres during secondary wall formation. Cytobiologie 6: 1-21

Roberts IN, Lloyd CW, Roberts K (1985) Ethylene-induced microtubule reorientations mediation by helical arrays. Planta 164: 439-447

Robinson DG, Quader H (1982) The microtubule-microfibril syndrome. In: Lloyd CW (ed) The cyto-skeleton in plant growth and development. Academic Press, London, pp 109-126

Robinson DG, Grimm I, Sachs H (1976) Colchicine and microfibril orientation. Protoplasma 89: 375-380

Roland JC, Mosiniak M (1983) On the twisting pattern texture and layering of the secondary cell walls of lime wood. Proposal of an unifying model. IAWA Bull 4: 15-26

Roland JC, Vian B (1979) The wall of the growing plant cell: its three-dimensional organization. Int Rev Cytol 61: 129-166

Roland JC, Reis D, Vian B, Satiat-Jeunemaitre B, Mosiniak M (1987) Morphogenesis of plant cell walls at the supermolecular level: internal geometry and versatility of helicoidal expression. Proto-plasma 140: 75-91

Saiki H (1970) Proportion of component layers in tracheid wall of early wood and late wood of some conifers. Mokuzai Gakkaishi 16: 244-249

Saiki H, Xu Y, Fujita M (1989) The fibrillar orientation and microscopic measurement of the fibril angles in young tracheids walls of sugi (*Cryptomeria japonica*). Mokuzai Gakkaishi 35: 786-792

Schneider B, Herth W (1986) Distribution of plasma membrane rosettes and kinetics of cellulose formation in xylem development of higher plants. Protoplasma 131: 142-152

Seagull RW (1991) Role of the cytoskeletal elements in organized wall microfibril deposition. In: Haigler CH, Weimer PJ (eds) Biosynthesis and biodegradation of cellulose. Marcel Dekker, New York, pp 143-163

Seagull RW (1992) A quantitative electron microscopic study of changes in microtubule arrays and wall microfibril orientation during in vitro cotton fibre development. J Cell Sci 101: 561-577

Shibaoka H (1991) Microtubules and the regulation of cell morphogenesis by plant hormones. In: Lloyd CW (ed) The cytoskeletal basis of plant growth and form. Academic Press, London, pp 159-168

Shibaoka H (1994) Plant hormone-induced changes in the orientation of cortical microtubules: altera-tions in the cross-linking between microtubules and the plasma membrane. Annu Rev Plant Physiol Plant Mol Biol 45: 527-544

Srivastava LM (1966) On the fine structure of the cambium of *Fraxinus americana* L. J Cell Biol 31: 79-93

Srivastava LM, O'Brien TP (1966) On the ultrastructure of the cambium and its vascular derivatives. 1. Cambium of *Pinus strobus* L. Protoplasma 61: 257-276

Srivastava LM, Sawhney VK, Bonnettemaker M (1977) Cell growth, wall deposition, and correlated fine structure of colchicine-treated lettuce hypocotyl cells. Can J Bot 55: 902-917

Taiz L (1984) Plant cell expansion: regulation of cell wall mechanical properties. Annu Rev Plant Physiol 35: 585-657

Takabe K, Fujita M, Harada H, Saiki H (1981a) The deposition of cell wall components in differentiating tracheids of sugi. Mokuzai Gakkaishi 27: 249-255

Takabe K, Fujita M, Harada H, Saiki H (1981b) Lignification process of Japanese black pine (*Pinus thunbergii* Parl.) tracheids. Mokuzai Gakkaishi 27: 813-820

Terashima N, Fukushima K, Sano Y, Takabe K (1988) Heterogeneity in formation of lignin. 5. Visualization of lignification process in differentiating xylem of pine by microautoradiography. Holzforschung 42: 347-350

Thomas RJ (1991) Wood: formation and morphology. In: Lewin M, Goldstein IS (eds) Wood Structure and Composition. Marcel Dekker, New York, pp 7-47

Timell TE (1986) Compression Wood in Gymnosperms. Springer, Berlin Heidelberg New York, pp 1-706

Tsoumis G (1991) Science and technology of wood: structure, properties, utilization. Van Nostrand Reinhold, New York, pp 1-494

Tsuda M (1975) The ultrastructure of the vascular cambium and its derivatives in coniferous species. 1. Cambial cells. Bull Tokyo Univ For 67: 158-226

Tuominen H, Sitbon F, Jacobsson C, Sandberg G, Olsson O, Sundberg B (1995) Altered growth and wood characteristics in transgenic hybrid aspen expressing *Agrobacterium tumefaciens* T-DNA indoleacetic acid-biosynthetic genes. Plant Physiol 109: 1179-1189

Tuominen H, Puech L, Fink S, Sundberg B (1997) A radial concentration gradient of indole-3-acetic acid is related to secondary xylem development in hybrid aspen. Plant Physiol 115: 577-585

Uehara K, Hogetsu T (1993) Arrangement of cortical microtubules during formation of bordered pit in the tracheids of *Taxus*. Protoplasma 172: 145-153

Uggla C, Moritz T, Sandberg G, Sundberg B (1996) Auxin as a positional signal in pattern formation in plants. Proc Natl Acad Sci USA 93: 9282-9286

Vesk PA, Rayns DG, Vesk M (1994) Imaging of plant microtubules with high resolution scanning electron microscopy. Protoplasma 182: 71-74

Vesk PA, Vesk M, Gunning BES (1996) Field emission scanning electron microscopy of microtubule arrays in higher plant cells. Protoplasma 195: 168-182

Vian B, Reis D (1991) Relationship of cellulose and other cell wall components: supramolecular organization. In: Haigler CH, Weimer PJ (eds) biosynthesis and biodegradation of cellulose. Marcel Dekker, New York, pp 25-50

Wardrop AB (1951) Cell wall organisation and the properties of the xylem. 1. Cell wall organisation and the variation of breaking load in tension of the xylem in conifer stems. Aust J Sci Res B-4: 391-414

Wardrop AB (1957) The organization and properties of the outer layer of the secondary wall in conifer tracheids. Holzforschung 11: 102-110

Wardrop AB (1958) The organization of the primary wall in differentiating conifer tracheids. Aust J Bot 6: 299-305

Wardrop AB (1964) The structure and formation of the cell wall in xylem. In: Zimmermann MH (ed) The formation of wood in forest trees. Academic Press, New York, pp 87-134

Wardrop AB, Dadswell HE (1950) The nature of reaction wood. 2. The cell wall organization of compression wood tracheids. Aust J Sci Res B-3: 1-13

Wardrop AB, Harada H (1965) The formation and structure of cell wall in fibres and tracheids. J Exp Bot 16: 356-371

Watanabe H, Tsutsumi J, Kojima K (1963) Studies of juvenile wood. 1. Experiments on stems of sugi trees (*Cryptomeria japonica* D. Don). Mokuzai Gakkaishi 9: 225-230

Wick SM, Seagull RW, Osborn M, Weber K, Gunning BES (1981) Immunofluorescence microscopy of organised microtubule arrays in structurally-stabilised meristematic plant cells. J Cell Biol 89: 685-690

Yamamoto H, Okuyama T, Yoshida M (1993) Method of determining the mean microfibril angle of wood over a wide range by the improved Cave's method. Mokuzai Gakkaishi 39: 375-381

Yoshizawa N (1987) Cambial responses to the stimulus of inclination and structural variation of compression wood tracheids in gymnosperms. Bull Utsunomiya Univ For 23: 23-141

Yuan M, Shaw PJ, Warn RM, Lloyd CW (1994) Dynamic reorientation of cortical microtubules, from transverse to longitudinal, in living plant cells. Proc Natl Acad Sci USA 91: 6050-6053

Yuan M, Warn RM, Shaw PJ, Lloyd CW (1995) Dynamic microtubules under the radial and outer tangential walls of microinjected pea epidermal cells observed by computer reconstruction. Plant J 7: 17-23

Zobel BJ, Jett JB (1995) Genetics of wood production. Springer. Berlin Heidelberg New York, pp 1-337

Zobel BJ, van Buijtenen JP (1989) Wood Variation: Its Causes and Control. Springer, Berlin Heidelberg New York, pp 1-363

4 Control of the Response to Biotic Stresses

Issei Kobayashi[1] and Yuhko Kobayashi[2]
[1]Laboratory of Plant Pathology, Faculty of Bioresources, Mie University, Mie, Japan
[2]Center for Molecular Biology and Genetics, Mie University, Mie, Japan

4.1
Summary

The cytoskeletal network of plant cells is a dynamic structure, changes in whose organization respond to external stimuli. An attack of pathogenic microbes represents an external stress that seriously threatens plant survival. Recently, it has been found that cytoskeletal elements, such as microtubules and microfilaments, are involved in plant defence reactions, especially in response to fungal penetration. Tubulin and actin inhibitors suppress the polarization of plant defence-related responses, such as massive cytoplasmic aggregation, deposition of papillae and accumulation of autofluorescent compounds at the sites of fungal penetration. Simultaneously, these inhibitors allow non-pathogenic fungi to penetrate successfully into non-host plants. Thus, microtubules and microfilaments, through the temporal and spatial regulation of molecules and/or organelles in the host cell, seem to control resistance responses against attempts of fungal penetration. On the other hand, plant cytoskeletal elements seem to play a critical role in the cell-to-cell spread of plant pathogenic viruses. In the tobacco mosaic virus, the movement protein P30 forms filaments that colocalize primarily with MTs. This association of P30 with cytoskeletal elements may play a critical role in intracellular transport of the P30-viral RNA complex through the cytoplasm to and possibly through plasmodesmata. These findings strongly suggest that the cytoskeleton plays a central role in both plant defence mechanisms and in the pathogenicity of microbes. The possibility of enhancing plant resistance to pathogens via an artificial manipulation of cytoskeletal elements will be discussed.

4.2
Significance of the plant cytoskeleton for pathogen resistance

The cytoskeleton, consisting of microfilaments and microtubules, is a highly conserved subcellular structure whose organization changes dramatically during cell cycle, development and adaptation. In plant cells, a dynamic reorganization of the cytoskeletal network has been observed during cell division (Katsuta and Shibaoka 1992) and cell wall synthesis (Hardham et al. 1980). The cytoskeleton can respond by adaptive reorganization to a variety of external stimuli such as gravitropic stimulation (Nick et al. 1991), low temperature (Quader et al. 1989), hormone treatment (Ishida and Katsumi 1991) and wounding (La Claire 1989) and this adaptive reorganization of the cytoskeleton in response to these abiotic stresses has been investigated in great detail. Biotic stresses, including pathogen attack, represent strong stimuli that pose serious threats to plant survival. However, little is known about the behaviour of the plant cytoskeleton and its possible function during the response to biotic stresses.

In mammals, phagocytosis of macrophages, which is important for host defence mechanisms as well as for tissue repair and morphogenetic remodelling, is driven by the reorganization of actin microfilaments (Caron and Hall 1998). Recently, a dynamic reorganization of cytoskeletal elements, such as microtubules and microfilaments, has been observed in several plant-fungus interactions, indicating that the cytoskeleton is involved in plant defence.

In contrast there is strong evidence that plant cytoskeletal elements may play a critical role in the pathogenicity of plant viruses. Recent studies show that the movement protein of the tobacco mosaic virus interacts with microtubules and microfilaments in vivo. This association of the movement protein with cytoskeletal elements suggests that the plant virus exploits and usurpates functions of the cytoskeletal network, in analogy to mammalian cells, where certain bacterial pathogens exploit the cytoskeleton of the host cell for internalization (reviewed in Higley and Way 1997; Dramsi and Cossart 1998).

A similar exploitation of the host cytoskeleton might occur during nodule formation induced in Fabaceae by symbiotic bacteria of the genus *Rhizobium*, where a reorganization of the plant cytoskeleton is thought to play an important role during nodule ontogeny. In this process, nodulation factors secreted by *Rhizobium* are possibly involved in the control of the cytoskeletal changes.

Altogether, these examples demonstrate the importance of cytoskeletal elements for both plant defence mechanisms and microbial pathogenicity and symbiosis. The first part of this chapter will therefore describe the dynamic reorganization of the cytoskeleton and discuss the possible role of the cytoskeleton during defence responses with focus on fungal pathogens. The second part deals with the role of the cytoskeleton during viral pathogenesis and bacterial symbiosis. The last part gives an outlook of the possibility of an enhancement of plant disease resistance through artificial manipulation of cytoskeletal elements.

4.3
Dynamic reorganization of the cytoskeleton during fungal attack

4.3.1
Reorganization of cytoskeletal network during attempts of fungal penetration

When plant cells perceive stimuli during fungal penetration attempts, visually detectable changes of cytoplasm are commonly observed including a movement of the nucleus towards the encounter site (Pappelis et al. 1974; Gross et al. 1993), cytoplasmic aggregation (Bushnell and Bergquist 1975; Kunoh et al. 1985), formation of cell wall appositions (papilla) beneath penetration sites (Aist 1976), alterations in the arrangement of cytoplasmic strands (Kitazawa et al. 1973; Kobayashi et al. 1993) and changes in the velocity of cytoplasmic streaming (Tomiyama 1956; Kobayashi et al. 1990). As motive force and track for the movements of organelles and cytoplasm the plant cytoskeleton is expected to play a role in these defence-related responses (Kamiya 1981; Williamson 1986; Seagull 1989).

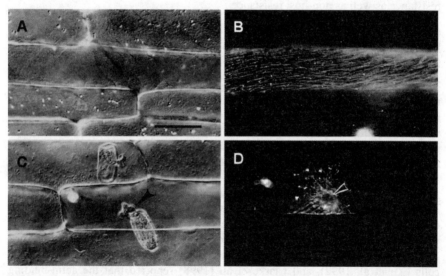

Fig. 4.1A-D. Rearrangement of microtubules in barley coleoptile cells inoculated with the non-pathogenic *E. pisi*. **A** and **C**: Differential interference contrast images. **D**: Epifluorescence images of immunostained microtubules. **A** and **B**: oblique orientation of microtubules in epidermal cells of uninoculated coleoptiles. **C** and **D**: microtubules 24 h after inoculation with numerous short and condensed microtubule bundles forming a radial array at the site of attempted penetration (**arrowheads**). **Bar** = 50 μm.

A dynamic reorganization of the cytoskeleton during attempted infection has been reported for several plant-fungus systems. In the interaction between barley coleoptiles and the non-pathogenic powdery mildew, *Erysiphe pisi*, microfilaments and microtubules are rearranged in the coleoptile during attempted penetration of the fungus (Kobayashi et al. 1991, 1992). When appressoria of the non-pathogenic fungus initiated penetration attempts, microfilaments and microtubules reorganized into a radial array directed towards the site of penetration (Fig. 4.1). In cultured parsley cells challenged by the non-pathogenic *Phytophthora infestans*, a localized depolymerization of microtubules occurred simultaneously with a rearrangement of microfilaments at the penetration site (Gross et al. 1993). Kobayashi et al. (1994) reported that such cytoskeletal rearrangements were characteristic for incompatible interactions between host and pathogen. In interactions between flax and flax rust fungus, *Melampsora lini*, radial arrays of microfilaments and microtubules at the encounter site were limited to incompatible interactions suggesting that this response followed the gene-for-gene rule. Conversely, major rearrangements of cytoskeletal components during interaction with pathogenic fungi have been observed in the cowpea and *Uromyces vignae* (Škalamera and Heath 1998) and the onion and *Botrytis allii* (McLusky et al. 1999) systems suggesting that the cytoskeleton is involved in host resistance as well as non-host resistance.

4.3.2
Possible role of the cytoskeleton in defence responses against fungal penetration

The actual role of the cytoskeleton during the manifestation of the non-host resistance is not yet elucidated. Generally, the cytoskeleton controls cell shape, cell motility, cell migration and cell polarity through the spatial and temporal regulation of molecules and organelles in the cell. In plants, the cytoskeleton participates in diverse cellular processes such as cytoplasmic streaming (Kamiya 1981), motility of organelles (Williamson 1986), cell wall synthesis (Hardham et al. 1980) and cell division (Katsuta and Shibaoka 1992). In addition to these well-known functions, the cytoskeleton has been discussed in many eukaryotic systems in the context of signal transduction pathways (reviewed in Tsukita et al. 1997; Gundersen and Cook 1999). Recent evidence supports this possibility for plants as well. PI 3-kinase, an important modulator protein for phosphatidylinositol (PI)-mediated signal transduction, is closely associated with the cytoskeleton in carrot cells (Xu et al. 1992), and Clarke et al. (1998) reported that the actin-binding protein profilin functions as stimulus-response modulator that translates signals into alterations of cytoplasmic architecture in pollen of *Papaver rhoeas*.

In this context, the following three possibilities can be perceived with respect to the potential role of the cytoskeleton in plant defence responses (Kobayashi et al. 1996):

1. Polarization of defence-related reactions.
2. Signal transduction for defence responses.

3. Cell-to-cell communication for spreading information between neighbouring cells.

Cytoplasmic aggregation and papilla formation are generally observed in plant cells that are subject of fungal penetration attempts. Ultrastructural study of cytoplasmic aggregates shows the presence of ER, Golgi apparatus and mitochondria beneath penetration sites (Bushnell and Zeyen 1976). The papilla that is formed in the centre of such cytoplasmic aggregates has been thought to represent an important defence reaction (Aist 1976; Heath and Heath 1971). In addition to the accumulation of organelles, it is observed that various defence-related compounds become localized around fungal penetration sites. In onion cells that have been inoculated with *Botrytis allii*, the fluorescent phenolic compound feruloyl-3-methoxytryamine accumulated at sites of attempted penetration. Localization of peroxidase and polarization of microfilaments at the same sites suggests that the polarization of microfilaments may contribute to the cross-linking of phenolic compounds into the cell wall through directed transport of the peroxidase. In barley cells, a similarly polar accumulation of protein, polysaccharide and fluorescent materials could be observed at fungal penetration sites, and this polar accumulation was completely blocked by treatment with inhibitors of cytoskeletal polymerization and depolymerization (Fig. 4.2, Kobayashi et al. 1997b). Simultaneously, treatment with these inhibitors effectively increased the penetration efficiency of the non-pathogenic fungus, *E. pisi*. *E. pisi* hardly succeeded in penetrating into barley, wheat, cucumber and tobacco cells in the absence of the inhibitors, whereas up to more than 60 % of appressoria could penetrate successfully into cells of these non-host plants and form haustoria upon treatment with cytochalasins (Kobayashi et al. 1997c). These results strongly suggest that a major role of the cytoskeleton in the defence reaction against fungal penetration attempts consists of a guided redistribution of defence-related compounds and organelles in attacked cells.

Several results provided evidence for an involvement of the plant cytoskeleton in the hypersensitive reaction. The hypersensitive reaction in barley coleoptile cells that were challenged by an incompatible race of *Erysiphe graminis hordei* was partially inhibited by the actin polymerization blocker cytochalasin B (Hazen and Bushnell 1983). Similarily, Škalamera and Heath (1998) reported that hypersensitive cell death in cowpea cells was inhibited by cytochalasin E during infection with the cowpea rust fungus (*Uromyces vignae*). In the flax and flax rust fungus (*Melampsora lini*) system, a rapid hypersensitive response developing about 24 h after inoculation normally inhibits fungal development and invasion in an incompatible interaction. However, in the presence of the microtubule polymerization blocker oryzalin, the occurrence of hypersensitive cell death was delayed and its frequency reduced (Kobayashi et al. 1997a).

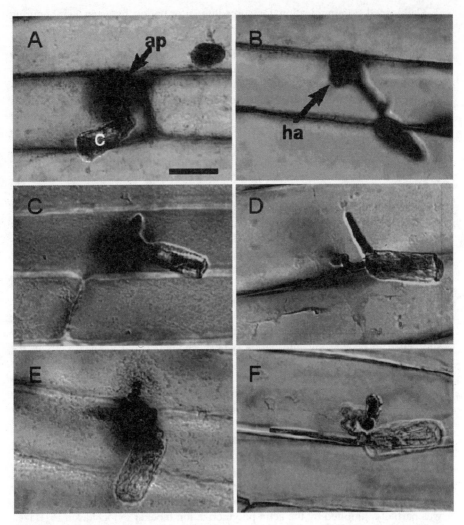

Fig. 4.2A-F. Bright-field micrographs showing the effects of the cytochalasin A on localized accumulation of defence-related materials in barley coleoptile cells 24 h after inoculation with *E. pisi*. **A, C, E** untreated controls. **B, D, F** treatment with 1 μgml⁻¹ cytochalasin A. **A,B** Signal after staining with amido black visualizing protein accumulation. **C,D** Signal after Acid Schiff reaction visualizing carbohydrate accumulation. **E,F** Signal after staining with esorcinol blue visualizing callose. Note the absence of signals and the successful penetration of *E. pisi* and formation of haustoria in the cytochalasin-treated cells shown in **B,D** and **F**. **ap** Appressorium; **ha** haustorium. **Bar** = 50 μm.

The exact function of the cytoskeleton in the expression of the hypersensitive response remains to be elucidated, but it is likely that the plant cytoskeleton at least partially regulates signal perception or transduction in the pathway leading to the hypersensistive response via binding of proteins that function as signal modulators (reviewed in Tsukita et al 1997; Zigmond 1996). In potato tissues treated with elicitors prepared from *Phytophthora infestans*, cytochalasin D and some inhibitors of the signal transduction cascade including staurosporine, ophiobolin and quinacrine inhibited the accumulation of rishitin, a potato phytoalexin, that is produced in potato tissues in response to the elicitor (Furuse et al. 1999). This result indicates that the factors involved in elicitor-induced signal transduction might be connected with actin cytoskeleton. Alternatively, the cytoskeleton might provide a rapid transport of these signals. In eukaryotic cells, external signals are received by receptors in or at the surface of cells, and the signals are transmitted to an intracellular target through a cascade of second-messenger modulation. Although the transmission of the signals is thought to be conveyed mainly by intracellular diffusion, the cytoskeleton provides tracks that might allow quick transmission of the modulated signals. Transfer of resistance from a cell that had been actually attacked by *E. pisi* to unchallenged adjacent cells was observed in barley (Kunoh et al. 1988). Recently, microfilaments have been detected as components of plasmodesmata in several plant species (White et al. 1994; Blackman and Overall 1998) raising the possibility that the cytoskeleton contributes to signal transmission between adjacent cells.

4.4
Involvement of the cytoskeleton in viral pathogenesis

To establish a systemic infection, plant viruses must move locally from cell to cell and enter the phloëm, through which they will move over long distances to establish a systemic infection and produce disease. A virus-encoded product, the movement protein (MP), actively potentiates viral cell-to-cell spread through plasmodesmata, the cytoplasmic bridges that function as intercellular connections (Gibbs 1976; Deom et al. 1992; Citovsky and Zambryski 1993; McLean et al. 1993; Lucas and Gilbertson 1994). In this chapter, we focus on the involvement of the cytoskeleton in cell-to-cell movement of viruses. There are a number of excellent recent reviews which described the general mechanism of viral movement in detail (Hull 1991; Lucas and Gilbertson 1994; Maule 1994; Carrington et al. 1996; Lartey and Citovsky 1997; McLean et al. 1997; Lazarowitz and Beachy 1999).

The 30-kDa protein (P30) MP encoded by tobacco mosaic virus (TMV) has been most extensively studied with respect to the mechanism of cell-to-cell movement. P30 can bind RNA in vitro and increase the size exclusion limits of plasmodesmata in mesophyll cells (Citovsky et al. 1990, 1992; Deom et al. 1992; Moore et al. 1992). Since White et al. (1994) demonstrated that microfilaments are associated with plasmodesmata, the possibility that the cytoskeleton could be involved in viral cell-to-cell movements in plants had been considered. Moreover, the cytoskeleton is known to act as a trafficking system for intracellular transport, translocation vesicles, organelles, protein, and even mRNA to specific cellular locations (Williamson 1986; Vale 1987; Dingwall 1992; Singer 1992; Wilhelm and Vale 1993; Bassell et al. 1994; Hesketh 1994). In animals, the cytoskeleton is involved in trafficking of parasite genomes to the nucleus and in the intracellular redistribution of viral proteins (Ben-Ze'ev et al. 1983; Pasick et al. 1994; Topp et al. 1994; Avalos et al. 1997; Li et al. 1998;). Therefore, it appeared plausible that associations between the P30-RNA complex and the cytoskeleton might be important in the spread of virus in plants.

Recently, this supposition was supported by the discovery that there are specific associations of P30 with microtubules and microfilaments during studies on the role of the cytoskeleton in directed transcytoplasmic movement and regulation of plasmodesmal function (Heinlein et al. 1995, 1998; McLean et al. 1995; Carrington et al. 1996;). Fusion proteins between jellyfish green fluorescent protein (GFP) and the MP coaligned with microtubules in infected tobacco protoplasts derived from the BY-2 cultured cell line (Heinlein et al. 1995 1998). Similarly, affinity-purified P30 polyclonal antibodies visualized a number of P30 filaments that were observed to colocalize with microtubules in both virus-infected and P30-transfected tobacco protoplasts (McLean et al. 1995). These coalignments were disrupted by low temperature (Heinlein et al. 1995; McLean et al. 1995) and by treatment with oryzalin and propizamide, which are known to disrupt microtubules, whereas no effect could be observed after treatment with the microtubule-stabilizing agent taxol (Heinlein et al. 1995). McLean et al. (1995) also examined a potential interaction between P30 and the actin cytoskeleton and could demonstrate that some of the P30 filaments colocalized with actin cytoskeleton. Furthermore, P30 appears to bind directly to actin and tubulin because P30 produced in overexpressors under control of the CaMV-35S promotor cosedimented with actin and tubulin in vitro. However, it is not known whether in vivo P30 interacts directly with the plant cytoskeleton or whether the binding to actin and tubulin is mediated by further proteins (McLean et al. 1995). Although treatments of cytochalasin showed a similar effect on the filamentous structure of P30, changes in the P30 network were much less pronounced than with cold treatment (McLean et al. 1995). Moreover, the colocalization between P30 filaments and microfilaments typically was less pronounced in number and extent than that seen between P30 filaments and microtubules (McLean et al. 1995). These results suggested that P30 protein associates mainly with microtubules and only partially with micro-filaments. A microtubule-binding activity could also be demonstrated in vitro for a further plant viral protein, the HSP70-related 65-kDa protein of the beet yellows closterivirus (Karasev et al. 1992).

A model for intracellular transport of P30 can be proposed based on the observed colocalization of P30 with cytoskeletal components and on the well-known fact that both microtubule motor and actin-myosin systems appear to actively transport various mRNAs as RNA-protein complexes in animal cells (Wilhelm and Vale 1993). Since viruses tend to exploit cellular mechanisms, the interaction of P30 with both microtubules and microfilaments may mimic the transport of RNA-protein complexes and organelles via microtubule motors and actin-myosin systems for long- and short-distance transport, respectively (Langford 1995).

The P30-RNA complex is predicted to be elongate and thin, because this would reduce diffusion within the cytoplasm while favouring organized or directed movement along cytoskeletal elements (Citovsky et al. 1992). According to their model, the cytoskeleton provides a track for the long unfolded P30-RNA complexes and facilitates linear, directed transport. P30 could be associated with the cytoskeleton either before, during, or after RNA complex formation. Then it would associate with microfilaments for short-distance unidirectional movement to and possibly through plasmodesmata because theses structures contain actin. Results of injection studies with fluorescent dextranes suggested that plasmodesmal gating could be controlled via microfilaments (Ding et al. 1996). Thus, P30 is proposed to interact with plasmodesmata-associated microfilaments usurpating them to be targeted and moved through plasmodesmata into the cytoplasm of adjacent cells. Since TMV-P30 can move between cells by itself (Waigmann and Zambryski 1995), P30 may be shuttled through plasmodesmata by the microfilaments extending between cells. The ability of P30 to increase the plasmodesmal size exclusion limit (Wolf et al. 1989, 1991; Waigmann et al. 1994) may also be related to its interaction with actin, because actin was found in the neck region of plasmodesmata, where the size exclusion limit is thought to be regulated (White et al. 1994). Summarizing, this evidence suggests that microtubules and filamentous actin may deliver complexes of MP with viral RNA to and through plasmodesmata (Heinlein et al. 1995; McLean et al. 1995; Carrington et al. 1996).

In plant cells, including BY-2 suspension cultured cells, microtubules are observed in the cortical cytoplasm at the cell periphery (cortical microtubules) and in association with the ER (Allen and Brown 1988; Hepler et al. 1990; Reuzeau et al. 1997). Interestingly, the association of MP with microtubules was most pronounced during the mid to late stages of infection and subsequent to its association with elements of the ER (Heinlein et al. 1998; Mas and Beachy 1998). It is unknown whether viral MPs harbour plasmodesmal targeting sequences. Based on analogies to nuclear import and the involvement of import receptors, such plasmodesmal targeting sequences are expected. Alternatively, viral MPs or viral replication complexes and MPs could form specific associations with subdomains of the cortical ER and/or cortical microtubules and microfilaments, which themselves are associated with plasmodesmata and act to guide the MPs toward these intercellular channels (Lazarowitz and Beachy 1999). In this case, microtubule and microfilament associated proteins, including molecular motors, and a functional characterization of the interaction between MPs, ER, and cytoskeleton becomes important (Lazarowitz and Beachy 1999).

4.5
Involvement of the cytoskeleton in rhizobial symbiosis

The symbiotic interactions between soil bacteria of the genera *Rhizobium*, *Azorhizobium* or *Bradyrhizobium*, which are referred to as rhizobia, and plants of the *Leguminosae* family result in the formation of nodules, new organs in which the bacteria reduce nitrogen into ammonia that can be subsequently utilized by the plant. The early stages of root nodule development are mediated by signal exchange between plant and rhizobia controlling altered gene expression on the bacterial part and cell growth, division and differentiation on the host part. In plants, cell shape and the direction of cell expansion depend on the correlation of cellulose microfibrils in the cell wall and plasma membrane-associated cortical microtubules with transverse microtubules maintaining cell elongation (see Chapt. 1; Giddings and Staehelin 1991; Williamson 1991). Similarly to actin microfilaments, the microtubular cytoskeleton changes its organization remarkably during cell division (Traas et al. 1987; Baluška and Barlow 1993;) and cell wall synthesis (Hardham et al. 1980; Seagull 1992; Goddard et al. 1994). Therefore, the cytoskeleton was expected to be involved in root nodule development.

Rhizobia produce Nod factors (NFs), whose synthesis is under the control of nodulation (nod) genes that are transcribed in the presence of plant flavonoids. NFs are signal molecules involved in most of the early developmental responses, in growth responses elicited by the corresponding bacteria, such as root hair induction and deformations, in the invasion of plant tissues by means of tubular structures called infection threads and in the formation of a nodule meristem whose activity ensures nodule (Newcomb 1981; Brewin 1991; Roth and Stacey 1989; Hirsch 1992; Kijne 1992; Long 1996).

The examination of cell division patterns provides the opportunity to explore the effect of NFs on plant development such as a so-called cytoplasmic activation that includes the formation of phragmosomes which was shown to be highly site specific and to occur early in response to NFs in some plants such as *Vicia*, (Van Brussel et al. 1992). The current hypothetical model proposes that NFs bind to plasmalemma-located receptors (Bono et al. 1995; Niebel et al. 1997), followed by subsequent signal transduction. In alfalfa root hairs, *Rhizobium meliloti* NFs induce a depolarization of plasma membrane potential (Ehrhardt et al. 1992; Felle et al.1995), cytoskeletal changes (Allen et al. 1994) and calcium spiking (Ehrhardt et al. 1996). It has been shown that *Rhizobium leguminosarum* bv. *trifolii* NFs are specifically internalized into clover root hairs (Philip-Hollingsworth et al. 1997).

Yang et al. (1994) found that the inner cortical cells that showed phragmosome formation were induced to enter the cell cycle via passage through the G_1 and S phases and continued to divide, forming nodule cortex and nodule meristem. In contrast, the outer cells that had been invaded by the bacteria ceased division. Therefore, it appears that rhizobia exploit cell division directly in the inner cell layers by inducing the formation of additional cells that will form the organ in which the bacteria will ultimately reside. In the outer cell layer, they seem to exploit, additionally, an indirect consequence of cell division, namely, increased cell wall synthesis and vesicular traffic. These cytoskeleton-dependent functions could help to render the usually inactive outer cortical cells as conductive to infection thread formation as are the tip-growing root hairs of the epidermis (Hirsch 1992; Kijne et al. 1992; Ridge 1992; Van Brussel et al. 1992; Yang et al. 1994).

The involvement of the plant cytoskeleton in early stages of nodulation has been suggested by studies demonstrating that cytoskeletal reorganizations occur at the tip of root hairs treated with NFs or in the root cortex of *Vicia hirsuta* that either had been infected by its specific symbiont or treated with NFs (Allen et al. 1994; Van Spronsen et al. 1995; Timmers et.al. 1998). Moreover, several of the symbiotic responses such as root hair induction and deformation, activation of cortical cells, oriented growth of an infection network in plant tissues, formation of nodulation-related division centres and cell enlargement (Truchet et al. 1991; Ridge 1992; Van Brussel et al. 1992; Ardourel et al. 1994; Yang et al. 1994) are dependent on the cytoskeleton. Thus, changes in cytoskeletal organization are likely to be involved in many of the symbiosis-related steps directing nodule development.

In the Leguminosae, the nodule meristem remains active for several weeks, thus leading to the formation of elongated indeterminate nodules comprising central and peripheral tissues. Timmers et al. (1998) provided experimental evidence for internalization of NFs and for architectural rearrangements of microtubules during nodule differentiation. In indeterminate nodules of alfalfa, the organization of microtubules in the central zone of the nodule changes tightly parallel to features of symbiotic differentiation that are related to cell infection, bacterial release, endopolyploidization, cell enlargement, spatial organization of cell components and organelle ultrastructure and positioning (Timmers et al. 1998). Their observations showed that in alfalfa nodules, these microtubular changes initiate in the nodule zone where rhizobial NFs are internalized in infected cells and that these changes strongly correlate with symbiosis-specific cell differentiation traits. Similar changes could be observed in different nodule types (Timmers et al. 1998),indicating that the link between cytoskeletal changes and rhizobial symbiosis might be general.

4.6
Prospects

Recent advances of plant molecular biology uncovered plant disease resistance genes at a molecular level. More than 20 resistance genes and their respective downstream genes have been cloned and their structure has been determined in detail (reviewed in Hammond-Kosack and Jones 1997). The defence-related signal cascade mediating race-specific cultivar resistance will be elucidated in the near future. Concerning applied science, this molecular understanding of defence responses should permit the development of disease-resistant crops with the help of molecular breeding techniques. As described above, the plant cytoskeleton plays an important role in plant defence mechanisms as well as viral pathogenesis and microbial symbiosis. As a highly conserved subcellular structure in higher plant cells, the cytoskeleton will provide new targets for molecular breeding of disease-resistant crops. Possible strategies to modify the cytoskeletal network in order to enhance plant disease resistance are as follows:

1. Regulation of cytoskeletal organization through modification of cytoskeletal regulators.
2. Control of expression of specific isotypes for tubulin (see Chapt. 7) or actin.
3. Structural modification of cytoskeletal proteins to change their affinity for other molecules.

Temporal and spatial control of defence responses appears to be highly important for an effective inhibition of pathogen attack. Little time is required for fungal pathogens to penetrate plant cell walls and to invade cells subsequently. In general, a hypersensitive reaction has to be triggered prior to a successful colonization by the pathogen in order to suppress the formation of lesions. Rapid responses of cytoskeletal network to microbial attack are therefore essential for a successful suppression of pathogen attack. The organization of the cytoskeleton is thought to be regulated by a number of binding proteins. Whereas such binding proteins have been isolated from yeast and animal cells, little is known about their plant counterparts. An exception is the actin-binding protein profilin that has been well characterized as a modulator of the actin cytoskeleton in both plant and animal cells (Sun et al. 1995; Steiger et al. 1997). Stable overexpression of birch pollen profilin in mammalian cells renders the actin network more resistant to depolymerizing agents (Rothkegel et al. 1996). This result indicates that the status of the cytoskeleton can be controlled through artificial regulation of cytoskeleton-associating proteins. Recently, small GTPases of the Rho family have emerged as key regulators of the actin cytoskeleton that appear to control coordinately different cellular activities through interaction with multiple target proteins (Hall 1998). Homologues of Rac, a member of the Rho GTPase family, have been cloned from several higher plants (Yang and Watson 1993; Borg et al. 1997; Winge et al. 1997). In fact, overexpression of a constitutively active (i.e. GTP-bound) form of a Rac gene results in an enhanced hypersensitive responses and in resistance to the blast disease in rice (Ono et al. 1999). Although it is not clear whether this effect was caused via a modified regulation of the actin cytoskeleton, such Rac GTPases are prime candidates for the manipulation of cytoskeletal reorganization.

In general, the different actin and tubulin isotypes are highly conserved in terms of amino acid sequence. However, the existence of multiple genes coding for these cytoskeletal proteins raises the possibility that different isotypes can have different roles in vivo (Meagher 1991; Chasan 1992). For example, the small *Arabidopsis* genome contains six expressed genes for α- and nine for β-tubulins (Kopczak et al. 1992; Snustad et al. 1992). It has been shown for animal cells that certain tubulin isotypes from the same animal are not functionally exchangable, although they are often interchangable with the same isotypes from different organisms (Ludueña 1993). The promoter of the maize *Tuba3* α-tubulin gene is activated by colonization with an arbuscular mycorrhizal fungus, whereas the closely related *Tuba1* promoter is not (Bonfante et al. 1996). The *Arabidopsis* *ACT7* actin gene was the only isotype to respond to several external stimuli including hormone treatments, light regime and wounding (McDowell et al. 1996). In barley leaves challenged by the non-pathogenic powdery mildew fungus, *E. pisi*, only one among three of the major expressed actin isotype genes was found to respond to the fungal attack (Hattori, K. and Kobayashi, I. unpubl. data) These results indicate that specific isotypes of cytoskeletal proteins are associated with plant defence responses.By controlling the expression of these specific isotypes artificially, one might venture to enhance disease resistance.

Changes in amino acid sequence of cytoskeletal proteins can cause changes in the affinity for other molecules. Anthony et al. (1998) reported that a point mutation of α-tubulin caused dinitroaniline herbicide resistance in plants, indicating that the mutation resides in the drug-binding site and that the change in the amino acid resulted in decrease of affinity between a-tubulin and herbicides. Modification of the domains where viral movement proteins interact with cytoskeletal proteins is expected to block cell-to-cell transmission of virus particles. Structural analysis of the binding partners and yeast two-hybrid approaches are expected to uncover the relevant domains and structures to develop this strategy.

Plants had to evolve a network of complex defence mechanisms consisting of diverse mechanisms designed to combat a wide variety of pathogens. The cytoskeleton controls a variety of cellular activities and thus appears to play an important role in some defence-related responses. Pathogens often take advantage of the genetic conservation and functional plasticity characteristic of the plant cytoskeleton. However, this means that the cytoskeleton is a good target for approaches with the aim of protecting plants from various biotic stresses.

Acknowledgements. This research was supported in part by a Grant-in-Aid for Encouragement of Young Scientists No.09760042 (1997) and a Grant-in-Aid for Scientific Research No.11660045 (1999) from the Ministry of Education, Science and Culture of Japan.

References

Aist JR (1976) Papillae and related wound plugs of plant cells. Annu Rev Phytopathol 14: 145-163

Allen NS, Brown DT (1988) Dynamics of the endoplasmic reticulum in living onion epidermal cells in relation to microtubules, microfilaments and intracellular particle movement. Cell Motil Cytoskeleton 10: 153-163

Allen NS, Bennet MN, Cox DN, Shipley A, Herhardt DW, Long SR (1994) Effects of Nod factors on alfalfa root hair Ca^{++} and H^+ currents and on cytoskeletal behavior. In: Daniels MJ, Downie JA, Osbourn AE (eds), advances in molecular genetics of plant-microbe interactions, vol 3. Kluwer, Dordrecht Boston London, pp 107-113

Anthony RG, Waldin TR, Ray JA, Bright SWJ, Hussey PJ (1998) Herbicide resistance caused by spontaneous mutation of the cytoskeletal protein tubulin. Nature 393: 260-263

Ardourel M, Demont N, Debellé F, Maillet F, de Billy F, Promé JC, Dénarié J, Truchet G (1994) *Rhizobium meliloti* lipooligosaccharide nodulation factors: different structural requirements for bacterial entry into target root hair cells and induction of plant symbiotic developmental responses. Plant Cell 6: 1357-1374

Avalos RT, Yu Z, Nayak DP (1997) Association of influenza virus NP and M1 proteins with cellular cytoskeletal elements in influenza virus-infected cells. J Virol 71: 2947-2958

Baluška F, Barlow PW (1993) The role of the microtubular cytoskeleton in determining nuclear chromatin structure and passage of maize root cells through the cell cycle. Eur J Cell Biol 61: 160-167

Bassell GJ, Taneja KL, Kislauskis EH, Sundell CL, Posers CM, Ross A, Singer RH (1994) Actin filaments and the spatial positioning of mRNAs. In: Estes JE, Higgins PJ (eds) Actin: biophysics, biochemistry and cell biology. Plenum Press, New York, pp 183-189

Ben-Ze'ev A, Abulafia R, Bratosin S (1983). Herpes simplex virus assembly and protein transport are associated with the cytoskeletal framework and the nuclear matrix in infected BSC-cells. Virology 129: 501-507

Blackman LM, Overall RL (1998) Immunolocalisation of the cytoskeleton to plasmodesmata of *Chara corallina*. Plant J 14: 733-741

Bonfante P, Bergero R, Uribe X, Romera C, Rigau J, Puigdomenech P (1996) Transcriptional activation of a maize α-tubulin gene in mycorrhizal maize and transgenic tobacco plants. Plant J 9: 737-743

Bono J-J, Riond J, Nicolaou KC, Bockovich NJ, Estevez VA, Cullimore JV, Ranjeva R (1995) Characterization of a binding site for chemically synthesized lipo-oligosaccharidic NodRm factors in particulate fractions prepared from roots. Plant J 7: 253-260

Borg S, Brandstrup B, Jensen TJ, Poulsen C (1997) Identification of new protein species among 33 different small GTP-binding proteins encoded by cDNAs from *Lotus japonicus*, and expression of corresponding mRNAs in developing root nodules. Plant J 11: 237-250

Brewin NJ (1991) Development of the legume root nodule. Annu Rev Cell Biol 7: 191-226

Bushnell WR, Bergquist S (1975) Aggregation of host cytoplasm and the formation of papillae and haustoria in powdery mildew of barley. Phytopathology 65: 310-318

Bushnell WR, Zeyen RJ (1976) Light and electron microscope studies of cytoplasmic aggregates formed in barley cells in response to *Erysiphe graminis*. Can J Bot 34: 1647-1655

Caron E, Hall A (1998) Identification of two distinct mechanisms of phagocytosis controled by different Rho GTPases. Science 282: 1717-1721

Carrington JC, Kasschau KD, Mahajan SK, Schaad MC (1996) Cell-to-cell and long-distance transport of viruses in plants. Plant Cell 8: 1669-1681

Chasan R (1992) Multitudinous microtubules. Plant Cell 4: 505-506

Citovsky V, Zambryski P (1993) Transport of nucleic acids through membrane channels: snaking through small holes. Annu Rev Microbiol 47: 167-197

Citovsky V, Knorr D, Schuster G, Zambryski P (1990). The P30 movement protein of tobacco mosaic virus is a single-stranded nucleic acid binding protein. Cell 60: 637-647

Citovsky V, Wong ML, Shaw AL, Venkataram Prasad, BV, Zambryski P (1992) Visualization and characterization of tobacco mosaic virus movement protein binding to single-stranded nucleic acid. Plant Cell 4: 397-411

Clarke SR, Staiger CJ, Gibbon BC, Franklin-Tong VE (1998) A potential signaling role for profilin of *Papaver rhoeas*. Plant Cell 10: 967-979

Deom CM, Lapidot M, Beachy RN (1992) Plant virus movement proteins. Cell 69: 221-224

Ding B, Kwon MO, Warnberg L (1996) Evidence that actin filaments are involved in controlling the permeability of plasmodesmata in tobacco mesophyll. Plant J 10: 157-164

Dingwall C (1992) Soluble factors and solid phases. Curr Biol 2: 503-505

Dramsi S, Cossart P (1998) Intracellular pathogens and the actin cytoskeleton. Annu Rev Cell Dev Biol 14: 137-166

Ehrhardt DW, Atkinson EM, Long SR (1992) Depolarization of alfalfa root hair membrane potential by *Rhizobium meliloti* Nod factors. Science 256: 998-1000

Ehrhardt DW, Wais R, Long SR (1996) Calcium spiking in plant root hairs responding to *Rhizobium* nodulation signals. Cell 85: 673-681

Felle HH, Kondorosi E, Kondorosi A (1995) Nod-signal induced plasma membrane potential changes in alfalfa root hairs are differentially sensitive to structural modifications of the lipochito-oligosaccharide. Plant J 7: 939-947

Furuse K, Takemoto D, Doke N, Kawakita K (1999) Involvement of actin filament association in hypersensitive reactions in potato cells. Physiol Mol Plant Pathol 54: 51-61

Gibbs AJ (1976) Viruses and plasmodesmata. In: Gunning BES, Robards AW (eds) Intercellular Communication In Plants: Studies on plasmodesmata, Springer Verlag, Berlin, pp 149-164

Giddings TH, Staehelin LA (1991) Microtubule mediated control of microfibril deposition: a reexamination of the hypothesis. In: Lloyd CW (ed) The cytoskeletal basis of plant growth and form, Academic Press, London, pp 85-99

Goddard RH, Wick SM, Silflow CD, Snustad DP (1994) Microtubule components of the plant cell cytoskeleton. Plant Physiol 104: 1-6

Gross P, Schmelzer, JCE, Hahlbrock K (1993) Translocation of cytoplasm and nucleus to fungal penetration sites is associated with depolymerization of microtubules and defence gene activation in infected, cultured parsley cells. EMBO J 12: 1735-1744

Gundersen GG, Cook TA (1999) Microtubules and signal transduction. Curr Opin Cell Biol 11: 81-94

Hall A (1998) Rho GTPases and the actin cytoskeleton. Science 279: 509-514

Hammond-Kosack KE, Jones JDG (1997) Plant disease resistance genes. Annu Rev Plant Physiol Plant Mol Biol 48: 575-607

Hardham AR, Green PB, Lang JM (1980) Reorganization of cortical microtubules and cellulose deposition during leaf formation in *Graptopetalum paraguayense*. Planta 149: 181-195

Hazen BE, Bushnell WR (1983) Inhibition of the hypersensitive reaction in barley to powdery mildew by heat shock and cytochalasin B. Physiol Plant Pathol 23: 421-438

Heath MC, Heath IB (1971) Ultrastructure of an immune and a susceptible reaction of cowpea leaves to rust infection. Physiol Plant Pathol 1: 277-287

Heinlein M, Epel BL, Padgett HS, Beachy RN (1995) Interaction of tobamovirus movement proteins with the plant cytoskeleton. Science 270: 1983-1985

Heinlein M, Padgett HS, Gens JS, Pickard BG, Casper SJ, Epel BL, Beachy RN (1998) Changing patterns of localization of the tobacco mosaic virus movement protein and replicase to the endoplasmic reticulum and microtubules during infection. Plant Cell 10: 1107-1120

Hepler PK, Palevitz BA, Lancelle SA, McCauley MM, Lichtscheidl I (1990) Cortical endoplasmic reticulum in plants. J Cell Sci 96: 355-373

Hesketh J (1994) Translation and the cytoskeleton: a mechanism for targeted protein synthesis. Mol Biol Rep 19: 233-243

Higley S, Way M (1997) Actin and cell pathogenesis. Curr Opin Cell Biol 9: 62-69

Hirsch AM (1992) Developmental biology of legume nodulation. New Phytol 122: 211-237

Hull R (1991) The movement of viruses within plants. Semin Virol 2: 89-95

Ishida K, Katsumi M (1991) Immunofluorescence microscopical observation of cortical microtubule arrangement as affected by gibberellin in *d5* mutant of *Zea mays* L.. Plant Cell Physiol 32: 409-417

Kamiya N (1981) Physical and chemical basis of cytoplasmic streaming. Annu Rev Plant Physiol 32: 205-236

Karasev AV, Kashina AS, Gelfand VI, Dolja VV (1992) HSP70-related 65 kDa protein of beet yellows clostervirus is a microtubule-binding protein. FEBS Lett 304: 12-14

Katsuta J, Shibaoka H (1992) Inhibition by kinase inhibitors of the development and the disappearance of the preprophase band of microtubules in tobacco BY-2 cells. J Cell Sci 103: 397-405

Kijne JW (1992) The *Rhizobium* infection process. In: Stacey G, Burris RH, Evans HJ (eds) Biological nitrogen fixation. Chapman and Hall, New York, pp 349-398

Kitazawa K, Inagaki H, Tomiyama K (1973) Cinephotomicrographic observations on the dynamic responses of protoplasm of a potato plant cell to infection by *Phytophtora infestans*. Phytopath Z 76: 80-86

Kobayashi I, Komura T, Sakamoto Y, Yamaoka N, Kunoh H (1990) Recognition of a pathogen and nonpathogen by barley coleoptile cells (I) Cytoplasmic responses to the nonpathogen, *Erysiphe pisi*, prior to its penetration. Physiol Mol Plant Pathol 37: 479-490

Kobayashi I, Kobayashi Y, Yamaoka N, Kunoh H (1991) An immunofluorescent cytochemical technique applying micromanipulation to detect microtubules in plant tissues inoculated with fungal spores. Can J Bot 69: 2634-2636

Kobayashi I, Kobayashi Y, Yamaoka N, Kunoh H (1992) Recognition of a pathogen and a nonpathogen by barley coleoptile cells (III) Responses of microtubules and actin filaments in barley coleoptile cells to penetration attempts. Can J Bot 70: 1815-1823

Kobayashi Y, Kobayashi I, Kunoh H (1993) Recognition of a pathogen and a nonpathogen by barley coleoptile cells. II. Alteration of cytoplasmic strands in coleoptile cells caused by the pathogen, *Erysiphe graminis*, and the nonpathogen, *E. pisi*, prior to their penetration. Physiol Mol Plant Pathol 43: 243-254

Kobayashi I, Kobayashi Y, Hardham AR (1994) Dynamic reorganization of microtubules and microfilaments in flax cells during the resistance response to flax rust infection Planta 195: 237-247

Kobayashi I, Kobayashi Y, Yamada M, Kunoh H (1996) The involvement of the cytoskeleton in the expression of non-host resistance in plants. In: Mills D, Kunoh H, Keen NT, Mayama S (eds) Molecular aspects of pathogenicity and host resistance: requirements for signal transduction. APS press, St. Paul, pp 185-195

Kobayashi I, Kobayashi Y, Hardham AR (1997a) Inhibition of rust-induced hypersensitive response in flax cells by the microtubule inhibitor oryzalin. Aust J Plant Physiol 24: 733-740

Kobayashi Y, Kobayashi I, Funaki Y, Fujimoto S, Takemoto T, Kunoh H (1997b) Dynamic reorganization of microfilaments and microtubules is necessary for the expression of non-host resistance in barley coleoptile cells. Plant J 11: 525-537

Kobayashi Y, Yamada M, Kobayashi I, Kunoh H (1997c) Actin Microfilaments are required for the expression of non-host resistance in higher plants. Plant Cell Physiol 38 725-733

Kopczak SD, Haas NA, Hussey PJ, Silflow CD, Snustad DP (1992) The small genome of *Arabidopsis* contains at least six expressed α-tubulin genes. Plant Cell 4: 539-547

Kunoh H, Aist JR, Hayashimoto A (1985) The occurrence of cytoplasmic aggregates induced by *Erysiphe pisi* in barley coleoptile cells before the host cell walls are penetrated. Physiol Plant Pathol 26: 199-207

Kunoh H, Katsuragawa N, Yamaoka N, Hayashimoto A (1988) Induced accessibility and enhanced inaccessibility at the cellular level in barley coleoptiles. III. Timing and localization of enhanced inaccessibility in a single coleoptile cell and its transfer to an adjacent cells. Physiol Mol Plant Pathol 33: 81-93

La Claire JW (1989) Actin cytoskeleton in intact and wounded coencytic green algae. Planta 177: 47-57

Langford GM (1995) Actin- and microtubule-dependent organelle motors: Interrelationships between the two motility systems. Curr Opin Cell Biol 7: 82-88

Lartey R, Citovsky V (1997) Nucleic acid transport in plant-pathogen interactions. Genet Engin 19: 201-214

Lazarowitz SG, Beachy RN (1999) Viral movement proteins as probes for intracellular and intercellular trafficking in plants. Plant Cell 11: 535-548

Li E, Stupack D, Bokoch GM, Nemerow GR (1998) Adenovirus endocytosis requires actin cytoskeleton reorganization mediated by Rho family GTPases. J Virol 72: 8806-8812

Long SR (1996) *Rhizobium* symbiosis: Nod factors in perspective. Plant Cell 8: 1885-1898

Lucas WJ, Gilbertson RL (1994) Plasmodesmata in relation to viral movement within leaf tissues. Annu Rev Phytopathol 32: 387-411

Ludueña RF (1993) Are tubulin isotypes functionally significant? Mol Biol Cell 4: 445-457

Mas P, Beachy RN (1998) Distribution of TMV movement protein in single living protoplasts immobilized in agarose. Plant J 15: 835-842

Maule AJ (1994) Plant-virus movement: De novo processing or redeployed machinery? Trends Microbiol 2: 305-306

McDowell JM, An YQ, Huang S, McKinney EC, Meagher RB (1996) The *Arabidopsis* ACT7 actin gene is expressed in rapidly developing tissues and responds to several external stimuli. Plant Physiol 111: 699-711

McLean BG, Waigmann E, Citovsky V, Zambryski PC (1993) Cell-to-cell movement of plant viruses. Trends Microbiol 1: 105-109

McLean BG, Zupan J, Zambryski PC (1995) Tobacco mosaic virus movement protein associates with the cytoskeleton in tobacco cells. Plant Cell 7: 2101-2114

McLean BG, Hempel FD, Zambryski PC (1997) Plant intercellular communication via plasmodesmata. Plant Cell 9: 1043-1054

McLusky SR, Bennett MH, Beale MH, Lewis MJ, Gaskin P, Mansfield JW (1999) Cell wall alterations and localized accumulation of feruloyl-3'-methoxytyramine in onion epidermis at sites of attempted penetration by *Botrytis allii* are associated with actin polarization, peroxidase activity and suppression of flavonoid biosynthesis. Plant J 17: 523-534

Meagher RB (1991) Divergence and differential expression of actin gene families in higher plants. Int Rev Cytol 125: 139-163

Moore PJ, Fenczik CA, Deom CM, Beachy RN (1992) Developmental changes in plasmodesmata In transgenic tobacco expressing the movement protein of tobacco mosaic virus. Protoplasma 170: 115-127

Newcomb W (1981) Nodule morphogenesis and differentiation. In: Giles KL, Atherly AG (eds) Biology of the Rhizobiacea. Int Rev Cytol, suppl 13. Academic Press, New York, pp 247-298

Nick P, Schäfer E, Hertel R, Furuya M (1991) On the putative role of microtubules in gravitropism of maize coleoptiles. Plant Cell Physiol 32: 873-880

Niebel A, Bono JJ, Ranjeva R, Cullimore JV (1997) Identification of a high affinity binding site for lipo-oligosaccharidic NodRm factors In the microsomal fraction of *Medicago* cell suspension cultures. Mol Plant-Microb Interact 10: 132-134

Ono E, Henmi K, Kawasaki T, Shimamoto K (1999) Enhanced disease resistance in transgenic rice expressing Rac genes. Plant Cell Physiol 40: 84

Pappelis AJ, Pappelis GA, Kulfinski FB (1974) Nuclear orientation in onion epidermal cells in relation to wounding and infection. Phytopathology 64: 1010-1012

Pasick JM, Kalicharran K, Dales S (1994) Distribution and trafficking of JHM coronavirus structural proteins and virions in primary neurrons and the OBL-21 neuronal cell line. J Virol 68: 2915-2928

Philip-Hollingsworth S, Dazzo FB, Hollingsworth R (1997) Structural requirements of *Rhizobium* chitolipooligosaccharides for uptake and bioactivity in legume roots as revealed by synthetic analogs and fluorescent probe. J Lipid Res 38: 1229-1241

Quader H, Hofmann A, Schnepf E (1989) Reorganization of the endoplasmic reticulum in epidermal cells of onion bulb scales after cold stress: involvement of cytoskeletal elements. Planta 177: 273-280

Reuzeau C, Doolittle KW, McNally JG, Pickard BG (1997) Covissualization in living onion cells of putative integrin, putative spectrin, actin, putative intermediate filaments, and other proteins at the cell membrane and in an endomembrane sheath. Protoplasma 199: 173-197

Ridge RW (1992) A model of legume root hair growth and *Rhizobium* infection. Symbiosis 14: 359-373

Roth LE, Stacey G (1989) Bacterium release into host cells of nitrogen-fixing soybean nodules: the symbiosome membrane comes from three sources. Eur J Cell Biol 49: 13-23

Rothkegel M, Mayboroda O, Rohde M, Wucherpfennig C, Valenta R, Jockusch BM (1996) Plant and animal profilins are functionally equivalent and stabilize microfilament in living animal cells. J Cell Sci 109: 83-90

Seagull RW (1989) The plant cytoskeleton. CRC Crit Rev Plant Sci 8: 131-167

Seagull RW (1992) A quantitative electron microscopic study of changes in microtubule arrays and wall microfibril orientation during in vitro cotton fibre development. J Cell Sci 101: 561-577

Singer RH (1992) The cytoskeleton and mRNA localization. Curr Opin Cell Biol 4: 15-19

Škalamera D, Heath MC (1998) Changes in the cytoskeleton accompanying infection-induced nuclear movements and the hypersensitive response in plant cells invaded by rust fungi. Plant J 16: 191-200

Snustad DP, Haas NA, Kopczak SD, Silflow CD (1992) The small genome of *Arabidopsis* contains at least nine expressed β-tubulin genes. Plant Cell 4: 549-556

Staiger CJ, Gibbons BC, Kovar DR, Zonia LE (1997) Profilin and actin depolymerizing factor: modulators of actin organization in plants. Trends Plant Sci 2: 275-281

Sun HQ, Kwiatkowska K, Yin HL (1995) Actin monomer binding proteins. Curr Opin Cell Biol 7: 102-110

Timmers AC, Auriac MC, de Billy F, Truchet G (1998) Nod factor internalization and microtubular cytoskeleton changes occur concomitantly during nodule differentiation in alfalfa. Development 125: 339-349

Tomiyama K (1956) Cell physiological studies on the resistance of potato plant to *Phytophthora infestans*. Ann Phytopathol Soc J 21: 54-62

Topp KS, Meade LB, LaVail JH (1994) Microtubule polarity in the peripheral processes of trigeminal ganglion cells: relevance for the retrograde transport of herpes simplex virus. J Neurosci 14: 318-325

Traas JA, Doonan JH, Rawlins DJ, Shaw PJ, Watts J, Lloyd CW (1987) An actin network is present in the cytoplasm throughout the cell cycle of carrot cells and associates with the dividing nucleus. J Cell Biol 105: 387-395

Truchet G (1978) Sur l'état diploide des cellules du meristme des nodules radiculaires des legumineuses. Ann Sci Nat Bot Biol Vég 19: 3-38

Truchet G, Roche P, Lerouge P, Vasse J, Camut S, De Billy F, Promé JC, Dénarié J (1991) Sulfated lipo-oligosaccharide signals of *Rhizobium meliloti* elicit root nodule organogenesis in alfalfa. Nature 351: 670-673

Tsukita S, Yonemura S, Tsukita S (1997) ERM (ezrin/radixin/moesin) family: from cytoskeleton to signal transduction. Curr Opin Cell Biol 9: 70-75

Vale RD (1987) Intracellular transport using microtubule-based motors. Annu Rev Cell Biol 3: 347-378

Van Brussel AAN, Bakhuizen R, Van Spronsen PC, Spaink HP, Tak T, Lugtenberg BJJ, Kijne W (1992) Induction of preinfection thread structures in the leguminous host plant by mitogenic lipooligosaccharides of *Rhizobium*. Science 254: 70-72

Van Spronsen PC, Van Brussel AAN, Kijne JW (1995) Nod factors produces by *Rhizobium leguminosarum* biovar *viciae* induce ethylene-related changes in root cortical cells of *Vicia sativa* ssp. *nigra*. Eur J Cell Biol 68: 463-469

Waigmann E, Zambryski P (1995)Tobacco mosaic virus movement protein-mediated protein transport between trichome cells. Plant Cell 7: 2069-2079

Waigmann E, Lucas WJ, Citovsky V, Zambryski P (1994) Direct functional assay for tobacco mosaic virus cell-to-cell movement protein and identification of a domain involved in increasing plasmodesmal permeability. Proc Natl Acad Sci USA 91: 1433-1437

White RG, Badelt K, Overall RL, Vesk M (1994) Actin associated with plasmodesmata. Protoplasma 180: 169-184

Wilhelm JE, Vale RD (1993) RNA on the move: the mRNA localization pathway. J Cell Biol 123: 269-274

Williamson R (1986) Organelle movements along actin filaments and microtubules. Plant Physiol 82: 631-634

Williamson RE (1991) . Orientation of cortical microtubules in interphase plant cells. Int Rev Cytol 129: 135-206

Winge P, Brembu T, Bones AM (1997) Cloning and characterization of rac-like cDNAs from *Arabidopsis thaliana*. Plant Mol Biol 35: 483-495

Wolf S, Deom CM, Beachy RN, Lucas WJ (1989) Movement protein of tobacco mosaic virus modifies plasmodesmatal size exclusion limit. Science 246: 377-379

Wolf S., Deom CM, Beachy RN, Lucas WJ (1991) Plasmodesmatal function is probed using transgenic tobacco plants that express a virus movement protein. Plant Cell 3: 593-604

Xu P, Lloyd CW, Staiger CJ, Drøbak BK (1992) Association of phosphatidylinositol 4-kinase with the plant cytoskeleton. Plant Cell 4: 941-951

Yang WC, de Blank C, Meskiene I, Hirt H, Bakker J, van Kammen A, Franssen H, Bisseling T (1994) *Rhizobium* Nod factors reactivate the cell cycle during infection and nodule primordium formation, but the cycle is only completed in primordium formation. Plant Cell 6: 1415-1426

Yang Z, Watson JC (1993) Molecular cloning and characterization of rho, a ras-related small GTP binding protein from garden pea. Proc Natl Acad Sci USA 90: 8732-8736

Zigmond SH (1996) Signal transduction and actin filament organization. Curr Opin Cell Biol 8: 66-73

5 Control of the Response to Aluminum Stress

Mayandi Sivaguru[1], Hideaki Matsumoto[1] and Walter J. Horst[2]
[1]Department of Cell Genetics, Research Institute for Bioresources, Okayama University, Chuo 2-20-1, Kurashiki 710-0046, Japan
[2]Institute of Plant Nutrition, University of Hannover, Herrenhäuserstr, 2, D-30419 Hannover, Germany

5.1
Summary

Soil acidity causes reduction in crop yield worldwide mainly caused by an aluminum (Al)-dependent inhibition of growth. Al acts primarily upon root growth. Despite the efforts of numerous studies, the physiological and molecular basis of this Al-induced inhibition of root elongation has remained unclear. Increasing evidence demonstrates the important role of the microtubular cytoskeleton in the control of cell division, cell elongation and the response to environmental stresses. Therefore, this chapter deals with the effect of Al on microtubular structure and function. After a short survey of cellular responses observed in Al-treated tissues, the cytoskeletal effects of Al treatment will be described with focus on Al-induced changes in microtubular structure in maize, wheat and tobacco cells. The chapter closes with a survey on the potential molecular mechanisms of Al toxicity and a perspective on genetic approaches to control Al toxicity.

5.2
Significance of aluminum toxicity for agriculture

Aluminum is, by abundance, the third element in the Earth crust and mainly bound in form of alumino-silicates. Al is released into forms that are available for plants as a consequence of soil acidification (pH < 5.5) caused by acid deposition, proton-producing processes in the soil and leaching of bases. Among the Al species in the soil solution Al^{3+} is the most important toxic species. However, one cannot exclude the toxic effect of monomeric and polymeric hydroxy-Al species (especially the Al-13 polymer, Kinraide 1993). Human activities have led to enhanced acidification of soils during the past few decades because of greater deposition of acid and acid-producing substances such as acid rain (Nouri and Reddy 1995), application of ammonium-based fertilizers (Foy 1988) and intensified cultivation of legumes (Bolan et al. 1991). It has been estimated that about 30% of the total ice-free land surfaces are composed of acid soils where the productivity of most food crops is severely restricted due to soil acidity (von Uexküll and Mutert 1995). Acid soils are commonly found in the humid temperate and especially in the tropical zones; but also in the semiarid tropics, acid soils are widespread (Table 5.1; Sanchez and Logan 1992).

Table 5.1. Main agroecological regions of the tropics suffering from constrainsts caused by Al toxicity (Sanchez and Logan 1992).

Problem	Semiarid tropics		Subhumid tropics (acid savannas)	
	Million ha	%	Million ha	%
Low nutrient reserves[a]	166	16	287	55
Al toxicity[b]	132	13	261	50
Acidity with Al toxicity[c]	298	29	264	50

[a]Less than 10% weatherable minerals in the sand-and-silt fraction. This constraint identifies highly weathered soils with limited capacity to supply P, K, C, Mg, and S. [b]More than 60% Al saturation in the top 50 cm. [c]Surface pH of less than 5.5 but less than 60% Al saturation.

In many tropical countries, amelioration of soil acidity and thus Al toxicity through liming is not possible owing to the prevailing socioeconomic conditions. To make a more productive use of such soils, therefore, plant species and cultivars with improved adaptation to acid soils and increased Al resistance have to be incorporated into the cropping system. Great differences exist in Al resistance between plant species, but also within species genotypic differences in Al resistance have been identified in most economically important food-crop species, and progress has been made in the breeding of high-yielding cultivars with increased Al resistance. Al-induced release of organic anions, especially citrate, malate (see reviews of Horst 1995; Kochian 1995; Rengel 1996) and oxalate (Ma et al. 1998), were found to represent important Al resistance mechanisms.

5.3
Al interaction with root growth

The initial effect of Al toxicity is the inhibition of root elongation, an effect occurring within minutes of Al application. It is generally accepted that the root apex plays the major role in Al perception and response (see Delhaize and Ryan 1995; Horst 1995; Kochian 1995; Taylor 1995; Rengel 1996; Matsumoto 2000 for recent reviews). This is well demonstrated by the following facts:

1. Al accumulation as an indicator of Al sensitivity takes place in the root apex (Delhaize et al. 1993a, Llugany et al. 1994; Sivaguru and Horst 1998);
2. Al-resistance mechanisms, such as the release of Al-complexing organic compounds, are confined mainly to the root apex (Horst et al. 1982; Delhaize et al. 1993b; Pellet et al. 1995);
3. Callose formation, a sensitive marker of Al sensitivity (Wissemeier et al. 1987; Zhang et al. 1994; Wissemeier and Horst 1995; Horst et al. 1997), is induced primarily in apical cells of the outer cortex (Sivaguru and Horst 1998).

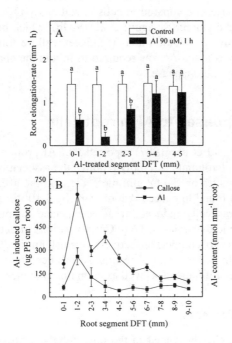

Fig. 5.1A,B.   Effect of aluminum (90 μM) administered in agarose blocks to intact roots of the Al-sensitive maize cultivar Lixis. **A** Effect on root-elongation rate after application of Al to individual 1-mm-zones at different distance from the root tip (**DFT**). **B** pattern of Al accumulation and Al-induced callose formation in individual 1-mm segments after application of Al to the entire root (apical 10 mm). In **A**, means with different letters are significantly different (P < 0.05, Tukey test) with regard to the Al effect. **Error bars** indicate SE (**A** and **B**) of five independent replicates (Sivaguru and Horst 1998; Horst et al. 1999).

However, the question of what is the primary target of Al within the root apex has remained open until recently. Ryan et al. (1993) could show that the root tip including the meristematic zone was the most Al-sensitive zone within the root. Sivaguru and Horst (1998) presented evidence that the distal part of the transition zone is the most Al-sensitive zone in primary roots of an Al-sensitive maize culti-var (Fig. 5.1A). The high sensitivity of this apical root zone as compared to other zones was related to an enhanced uptake of Al into this zone and expressed by enhanced callose formation independently of whether Al was applied to the entire root apex or to confined regions of the root apex (Fig. 5.1B). Recent evidence also suggests that differences in Al resistance between different maize cultivars are specifically expressed in this apical root zone (Kollmeier et al. 2000). In contrast to the transition zone, the elongation zone itself, where the major part of root elongation takes place (Evans and Ishikawa 1997, Kollmeier et al. 2000) was surprisingly insensitive to Al (Ryan et al. 1993, Sivaguru and Horst 1998; Koll-meier et al. 2000). In agreement with these results, in intact maize roots we could show that tobacco cells in suspension culture were more Al-sensitive during the logarithmic division phase as compared to cells in stationary phase that were expanding, but not longer dividing (Sivaguru et al. 1999b), which became manifest as weaker extent of growth inhibition and Al-induced callose formation. During stationary phase, twice as much Al was required to mimic the changes that were observed during the logarithmic phase.

5.4
Cytoskeletal responses to Al in root cells

A decade ago, MacDonald et al. (1987) reported that Al promoted microtubule assembly in vitro. As microtubule assembly requires the association of Mg^{2+} with GTP and GDP binding sites on the tubulin molecule, and the affinity of Al^{3+} for these binding sites was found to exceed that of Mg^{2+} by $3 \ 10^7$ times, the hydrolysis of GTP became drastically reduced, what is expected to interfere with the regulation of microtubule dynamics in vivo. Concentrations of active Al^{3+} as low as $4 \ 10^{-10}$ M were able to compete effectively against millimolar activities of Mg^{2+}. Consistent with these biochemical results, Al has been found to exert drastic effects upon actin microfilaments as well as upon microtubules in plant roots.

5.4.1
Al and actin microfilaments

Root morphogenesis is closely related to the microtubular cytoskeleton (Barlow and Parker 1996), while the onset of root-cell elongation is thought to be acto-myosin-dependent (Baluška et al. 1997). With respect to Al interaction with the cytoskeleton in vivo, Alfano et al. (1993) studied the long-term effect of Al on the actin MFs in *Riccia fluitans*, and Grabski and Schindler (1995), using a novel cell optical displacement technique, reported that exposure of plant cells to Al increased the stability of actin MFs in suspension-cultured soybean cells. They showed that this Al-induced rigour or tension within the actin network is concentration-dependent. In addition, a specific, dose-dependent relationship between Al and the protective effect of NaF and Mg^{2+} on actin filaments was demonstrated in these soybean suspension-culture cells. Similarily, in intact maize roots, effects of Al on the actin cytoskeleton in the elongation zone (Blancaflor et al. 1998) as well as in the transition zone (Sivaguru et al. 1999a) have been described. However, these effects became manifest only after prolonged Al treatment and therefore might represent the consequence rather than the cause of the rapid Al-induced inhibition of root elongation.

5.4.2
Al and microtubules

Considerable efforts have been directed to understanding the relation between Al-induced neurological defects in human and animal cells and aberrations of tubulin assembly and dysfunctions of microtubule-related proteins (Schmidt et al. 1991 and references therein). In intact plant roots, the reports available so far indicate an Al-induced depolymerization of cortical microtubules in wheat roots (Sasaki et al. 1997), an Al-induced stabilization of cortical microtubules in the elongation zone (Blancaflor et al. 1998) and a rapid disintegration of cortical microtubules in the transition zone (Horst et al. 1999; Sivaguru et al 1999a) of maize roots. Blancaflor et al. (1998) reported an initial stabilisation (cells of the outer cortex) and a reorientation (cells of the inner cortex) of cortical microtubules in the elongation zone of the maize root apex. In their detailed study, they also pointed out the initial stabilization of cortical microtubules (after 3 h of Al treatment), closely correlating with the simultaneous growth inhibition observed after Al treatment in the elongation zone at 3-4 mm distance from root tip (see Fig. 1C of Blancaflor et al. 1998). On the other hand, in an analysis of the distal transition zone 1-2 mm distant from root tip, which was identified as the most Al-sensitive apical root zone of maize (Fig. 1b of Sivaguru and Horst 1998), we found a depolymerization of cortical microtubules in the epidermal and the outermost cortical layer as early as 1 h after incubation with Al (Fig. 5.2), while a stabilization of cortical microtubules (as observed by Blancaflor et al. 1998) or no gross alterations to their structure were observed in the more proximal elongation zone of the same root apex (Sivaguru et al. 1999a).

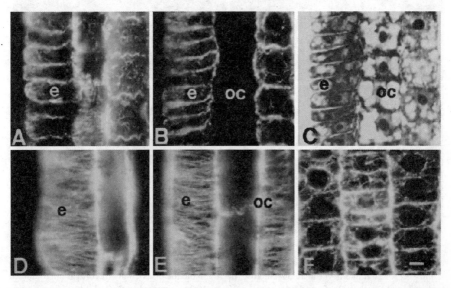

Fig. 5.2A-F. Effects of short-term Al treatment (90 μM, 1 h) on microtubules in cells of the distal transition zone (**A-C**) and of the elongation zone (**D-E**) of maize roots. **A,D** Control. **C** DIC image corresponding to the image in **B**. The most sensitive cells with respect to Al effects on the microtubular cytoskeleton proved to be the outermost cortical cells of the distal transition zone, as these lost all of their microtubules within 1 h of Al treatment (**B**). In contrast, no effects were found for the microtubules in the elongation zone (**D,E**). Extremely dense endoplasmic microtubules were induced by 1 h of Al in the apical meristem (**F**). E Epidermis; oc outer cortex. **Bar** 8 μm for **A-E** and 6 μm for **F** (Sivaguru et al. 1999a).

Both studies are consistent in the light of recent observations made in tobacco suspension cultures: during the logarithmic phase of the culture when cell division persists and elongation starts, comparable to the situation in the distal transition zone of the maize root apex (Baluška et al. 1996), Al caused a significant growth inhibition (Sivaguru and Horst 1998) accompanied by a depolymerization of cortical microtubules. In contrast, the cells in the stationary phase that have completed active growth and are no longer cycling (comparable to the situation in the elongation zone of intact maize root apex) showed an Al-induced stabilization (Fig. 5.3). These findings show that Al alters the microtubular cytoskeleton in a time- and concentration-dependent manner but depending on inherent differences that are related to the growth and developmental phase of the target cells. These differences might be related to differences either in the composition of tubulin isotypes or in the pattern of associated proteins, as suggested for the microtubular response to chilling stress in tobacco cells (Mizuno 1992 and Chap. 6, this book).

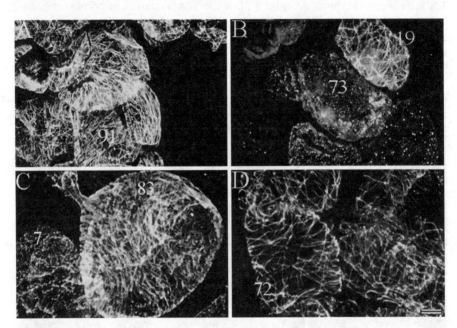

Fig.5.3A-D. Immunofluorescence images of tobacco cells (var. Samsun) in suspension culture during the logarithmic phase (**A,B**) and the stationary phase (**C,D**) before (**A,C**) and after (**B,D**) treatment with 50 µM Al for 24 h in 3 mM CaCl₂, 3% Sucrose, pH 4.5. Note the disassembly of cortical microtubules induced by Al in the cells from the logarithmic phase (**B**) as compared to the stability of cortical microtubules against Al in cells from the stationary phase (**D**). **Bar** 10 µm. (Sivaguru et al. 1999b).

5.5
Interaction of Al with the cytoskeleton – potential mechanisms

5.5.1
Apoplastic effects of Al

Externally applied Al rapidly binds to root cap mucilage (Archambault et al. 1996), which may protect the root meristem (Horst et al. 1982), and to the cell walls of root cells (Zhang and Taylor 1989; Blamey et al. 1990; Delhaize et al. 1993a), where the main binding sites are the negatively charged carboxylic groups of the pectic matrix and glucoprotein chains extending from the plasma membrane (Pettersson and Strid 1989). These charges yield an electrical potential gradient determining binding and distribution of ions in the apoplast (Kinraide 1993). Horst et al. (1999) showed that the spatial Al sensitivity of the root apex is positively correlated to the pectin contents of the root zones with the exception of the meristematic zone (which might have been due to contamination of this zone with mucilage). The importance of pectin content and the degree of pectin methylation defining the negative charge, for Al toxicity and resistance was further confirmed by Schmohl, N. and Horst, W.J. (submitted). Uptake of Al into the symplast has been demonstrated using different techniques, uptake kinetics (Zhang and Taylor 1990), staining of Al using fluorescent dyes (Tice et al. 1992), and microlocalization using EDX (Vasquez et al. 1999), LAMMA (Marienfeld et al. 2000) and SIMS (Lazof et al. 1994). Furthermore, studies with giant algae (Reid et al. 1995) confirmed clearly that Al enters the symplast. However, the calculated Al fluxes across the plasma membrane were extremely small (Rengel 1996). The question of whether Al-induced inhibition of root elongation is primarily due to apoplastic or to symplastic lesions is subject of an ongoing debate (Horst 1995; Kochian 1995, Rengel 1996; Vasquez et al. 1999; Matsumoto 2000). However, in our opinion there is little doubt that the primary Al injury in roots can be fully explained on the basis of apoplastic Al lesions.

5.5.2
Interaction of Al with cellular functions involved in microtubule structure and dynamics

Al induces the formation of callose ($1,3$-β-D-glucan), a specific marker of Al stress (Wissemeier et al. 1987) in a genotype-dependent manner (Horst et al. 1997). The intensity of callose induction was found to be tissue-specific in primary roots of maize and confined to the outer one or two layers of the apical root cortex (Sivaguru and Horst 1998, Sivaguru et al. 1999a). It should be emphasized that these outer cortical cells act as "master cells" and control the response to various environmental stimuli including gravitropism (Baluška and Hasenstein 1997 and references therein). The induction of callose synthesis points to modifications of plasma-membrane structure, function, altered cell-wall configuration and to increases of cytosolic [Ca^{2+}] levels (Kauss 1996). Consistently with these phenomena accompanying the induction of callose synthesis, Lindberg and Strid (1997) and Zhang et al. (1998) reported that Al treatment caused an increase in cytosolic $[Ca^{2+}]$ in isolated wheat protoplasts as well as in the intact wheat root apices, respectively. Alterations of plasma-membrane structure and function are also indicated by changes in cell surface electrical potential (Kinraide et al. 1992; Papernik and Kochian 1997; Takabatake and Shimmen 1997, Sivaguru et al. 1999a) and modified patterns of membrane lipids (Zhang et al. 1997) by Al (see also Horst 1995; Kochian 1995; Matsumoto 2000). Al apparently interacts directly and/or indirectly with factors that influence the organization of the cytoskeleton such as levels of cytosolic Ca^{2+} (as above and e.g. Jones et al. 1998a,b), Mg^{2+} and calmodulin (Haug 1984; Grabski et al. 1998). Alterations of cytosolic calcium levels induced by signalling substances such as auxin resulted in increased rigidity of the actin cytoskeleton in soybean root cells (Grabski and Schindler 1996). The latest work from their laboratory revealed that Al affects actin microfilaments via the involvement of calcium-regulated kinases and phosphatases (Grabski et al. 1998). Pertinent with this the microtubular cytoskeleton became disorganized by inhibitors of protein kinases and phosphatases (e.g. calyculin A and staurosporine). This was specifically observed for cortical microtubules, leading to impaired root elongation through induction of lateral growth and swelling in *Arabidopsis* (Baskin and Wilson 1997). Furthermore, swelling of Al-treated cells or tissues is often associated with a cessation of both root and root-hair elongation (Jones et al. 1995). Recently Jones et al. (1998b) were able to show an Al-induced increase in the cytosolic calcium levels in *Arabidopsis* root hairs.

Thus, some of the major factors influencing structure and function of microtubular and actin cytoskeleton are affected by Al (Giddings and Staehelin 1991; Simmonds 1992; Shibaoka 1994), and Al can interact with specific cell processes that require dynamic tubulin and actin cables such as differentiation, cell-plate formation, chromosome movement, cell-wall biosynthesis, nuclear and vesicular migration and secretion (Gunning and Hardham 1982; Lloyd 1991

5.5.3
Action of Al through the cell wall/plasma membrane/cytoskeletal continuum

Our understanding of cell wall and cytoskeletal physiology, and its dynamic role in cell elongation, has increased significantly in the recent past. For instance, the physical interaction between the microtubules and cellulose synthesis complexes through the plasma membrane (Giddings and Staehelin 1991, Fischer and Cyr 1998), the microtubular control of deposition and orientation of newly synthesized cellulose microfibrils, the interaction of growth regulators on microtubules and microfibrils and their interaction during cell elongation have been analyzed in great detail (Williamson 1991; see also Chap. 1). However, surprisingly, the potential interactions of Al with these components have not been investigated so far.

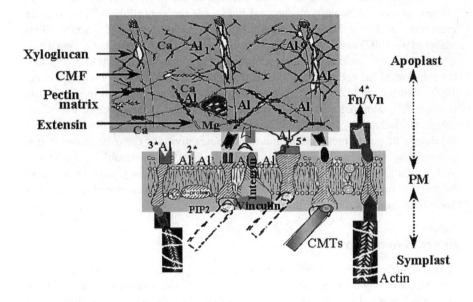

Fig. 5.4. A hypothetical and schematic description of the effect of Al on the cell wall/plasma membrane/cytoskeletal continuum. Under Al-free conditions the load bearing xyloglucans (**XG**) separate due to strain, and simultaneously the alternating cables of cortical microtubules (**CMTs**) mutually slide and pave the way to cell elongation (Cyr 1994). The model attempts to illustrate the point that apoplastic action of Al would be sufficient to effectively disturb the structural integrity of the underlying cytoskeleton via different mechanisms numbered 1* through 5*. 1* Al binding to calcium-binding sites especially with the pectic matrix and other negatively charged polymers may restrict and/or limit the breakdown of hemicelluloses which may lead to a decrease in wall loosening. 2* Al binding to the negatively charged lipids and protein moieties on the external face of the plasma membrane (PM) leads to membrane depolarisation thereby affecting the underlying membrane-bound cytoskeleton. 3* Al binding to the receptor sites of the plasma membrane may lead to increase in the cytosolic Ca^{2+} levels thereby inducing disassembly of cortical microtubules and actin microfilaments. 4* Al may also bind to other unknown extracellular matrix or adhesion proteins like fibronectin/vitronectin, thus uncoupling the information flow between cytoskeleton and cell-wall continuum. 5* Al binding to pectic and other extracellular polymer matrices may result in transmission of mechanical stress through the plasma membrane thus triggering alterations of the underlying cytoskeletal units. Details of the interaction of Al with the components of this model are given in the text (for additional information see Horst 1995). The molecular components of this model have not been identified and characterized in plant cells. **CMF** Cellulose microfibrils; **PM** plasma membrane, **CMTs** cortical microtubules. (After Wyatt and Carpita 1993, Horst 1995 and Cyr 1994).

The binding of Al to the pectin matrix and to other components of the cell wall is expected to produce mechanical strains due to a decreased extensibility of the cell wall, and these strains could be transduced from the external face of the plasma membrane through mechanical links between exocellular matrix, plasma membrane and cytoplasm dynamics (see Figure 3 of Wyatt and Carpita 1993; Miller et al. 1997) to the cytoskeleton, imposing detrimental effects on cytoskeletal structure and assembly. This assumption is supported by reports showing that the removal of the cell wall caused a disruption of microtubule arrays (Simmonds 1992). Consistent with this, recent work in intact maize plants and in maize suspension cells demonstrated that Al sensitivity (manifest as inhibition of root elongation and induction of callose synthesis) can be modulated by pectin content and the degree of pectin methylation and thus the negativity of the exocellular matrix determining Al binding to the apoplast (Horst et al. 1999, Schmohl, N. and Horst, W.J. submitted). A model depicting this possible effect of apoplastic Al on transmembrane signaling and its effect on the dynamics of microtubules and actin microfilaments is shown in Fig. 5.4.

5.5.4
Effect of Al on auxin transport

Recently, Ruegger et al. (1997) showed that root growth in the *Arabidopsis* mutant *tir3*, that has been isolated by its resistance to the auxin-transport inhibitor NPA (N-1-napthylethylpthalamic acid) and exhibits a reduction of about 50% of NPA binding sites is less sensitive to inhibition by NPA by about 20-50-fold as compared to the wild type. As a consequence, the extent of morphological disorders such as induction of periclinal divisions in the cortical cells and lateral cell expansion typically produced by prolonged treatment with auxin-transport inhibitors was found to be much less pronounced in the mutant as compared to the wild type.

Al can induce similar disorders, such as swelling of the outer cortical cells and periclinal divisions in Al-sensitive cultivars of maize or wheat (Blancaflor et al. 1998; Sivaguru et al. 1999a; Vasquez et al. 1999). Anomalous periclinal divisions at the root meristem-root-cap boundary (Sivaguru et al. 1999a) could be observed for prolonged treatment with both Al and NPA, consistent with data on maize roots published for the auxin-transport inhibitor TIBA (1,3,5-triiodobenzoic acid) (Kerk and Feldman 1994).

The similarity between the morphological effects produced by Al and those caused by auxin-transport inhibitors indicates that Al might inhibit root growth via changing auxin efflux possibly by interacting with NPA-binding proteins. Such a NPA-binding protein (NPB) has been identified recently and found to be an integral protein of the PM (Bernasconi et al. 1996). Additionally, NPA binding has been discussed to be associated with the cytoskeleton (Cox and Muday 1994).

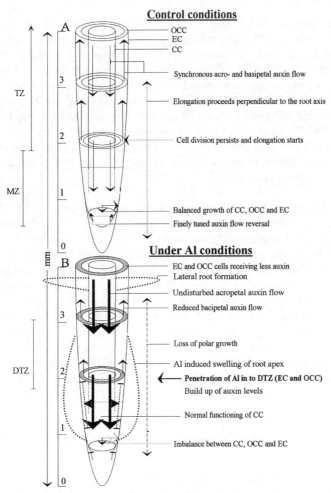

Control conditions

OCC
EC
CC

Synchronous acro- and basipetal auxin flow

Elongation proceeds perpendicular to the root axis

Cell division persists and elongation starts

Balanced growth of CC, OCC and EC
Finely tuned auxin flow reversal

Under Al conditions

EC and OCC cells receiving less auxin
Lateral root formation

Undisturbed acropetal auxin flow
Reduced bacipetal auxin flow

Loss of polar growth

Al induced swelling of root apex
Penetration of Al in to DTZ (EC and OCC)
Build up of auxin levels

Normal functioning of CC

Imbalance between CC, OCC and EC

Fig. 5.5A,B. A model depicting the effects of Al on auxin transport in the root apex. **A** Situation under control conditions without Al supply. **B** Proposed alterations in auxin transport, root growth and morphology in presence of Al. Enhanced uptake of Al into the epidermal and outer cortical cells in the distal transition zone (**DTZ**) compared to the meristematic zone (**MZ**) and the elongation zone (**EZ**) leads to blockage of basipetal auxin flow. Since acropetal auxin flow through the central cylinder is not affected by Al, this leads to an initial increase in the levels of auxin in the meristematic and distal transition zones, which may directly alter microtubular organization and dynamics and induce microtubule-related morphological changes such as periclinal divisions and root-tip swelling. Ultimately, this inhibition of basipetal auxin flow may also lead to an auxin accumulation in the central cylinder (**CC**), which could explain the elevated formation of lateral roots commonly observed in presence of Al. Inhibition of basipetal auxin flow in the distal transition zone also leads to auxin depletion in the cortical cells of the elongation zone, resulting in inhibition of cell and thus root elongation. The illustration suggests a signalling pathway in the root apex mediating the Al signal between distal transition zone and elongation zone through basipetal auxin transport. **OCC** Outer cortical cells; **EC** epidermal cylinder; **CC** central cylinder; **CEZ** central elongation zone; **DTZ** distal part of the transition zone (Sivaguru and Horst 1998). (After Hasenstein and Evans 1988, Müller et al. 1998 and Kollmeier et al. 2000).

It has been clearly established that coordinated auxin transport is involved in the regulation of root growth, morphology and the gravitropic growth response (Hasenstein and Evans 1988; Kaufman et al. 1995; Evans and Ishikawa 1997; Ruegger et al. 1997; Müller et al. 1998). Auxin is transported from auxin synthesizing shoot tissues via the phloëm towards the root apical meristem where it is proposed to be unloaded from the central stele into cortical and epidermal cells and then translocated basipetally to the elongation zone (Hasenstein and Evans 1988; Estelle 1998). This model has been strongly supported by the application of auxin transport inhibitors (see above) and more recently by molecular tools using *Arabidopsis* mutants defect in auxin-binding (Ruegger et al. 1997) and auxin-transport proteins (Müller et al. 1998).

Recently, Kollmeier et al. (2000) could show that the application of Al, TIBA or NPA to the transition zone led to a reduced [^{3}H]-IAA transport into the elongation zone while the IAA contents relative to control roots became elevated in the more distal transition zone and the meristematic zone. The relation between Al-induced inhibition of root growth and a possible interaction between Al and auxin transport was further supported by the result that exogenous application of IAA in root growth-stimulating doses to the elongation zone could alleviate the inhibition of root elongation caused by Al or TIBA administered to the transition zone. Interestingly, the inhibition of [^{3}H]-IAA transport by Al was more pronounced in an Al-sensitive as compared to an Al-resistant cultivar (Kollmeier et al. 2000). These findings support previous results (Hasenstein and Evans 1988) and suggest that an inhibition of auxin-transport is involved in Al toxicity (Fig. 5.5), although the mechanism by which Al affects basipetal IAA transport in the root needs to be further elucidated.

To date, five principle ways of explaining the Al-effect on basipetal IAA transport seem to be possible:

1.   Direct impact of Al on the IAA efflux carrier and/or a receptor protein in the plasma membrane of cortical cells regulating basipetal auxin flow as is the case for the inhibitors TIBA and NPA (Ruegger et al. 1997, and Miller et al., 1997).
2.   Modification of the cytoskeleton by Al as demonstrated by Blancaflor et al. (1998) in the elongation zone and by Sivaguru et al. (1999a) primarily in the distal transition zone which might interfere with the auxin transport system (Blancaflor and Hasenstein 1995). However, recently, Hasenstein et al. (1999) presented evidence that the microtubular cytoskeleton does not interact with auxin transport.
3.   Al activation of oxidative catabolism of IAA via peroxidases (Lagrimini et al. 1997).
4.   Al-induced changes in Ca^{2+} homeostasis (see above) which might affect auxin flow (Young and Evans 1994).
5.   Aluminum-induced callose at the plasma-membrane phase of cross walls in root epidermis and outer cortex (maize: Sivaguru et al. 1999a; *Arabidopsis* and wheat: unpubl. own results) that may act as a physical barrier and inhibit the basipetal transport of IAA. The auxin efflux carriers that are involved in root gravi-

tropism are located especially at these membranes of epidermal and outer cortical cells (see Figs. 7 and 8 of Müller et al. 1998).

5.6
Perspectives: genetic aspects of Al tolerance

Genes that confer resistance to Al have been investigated in *Arabidopsis*, yeast and wheat (see Kochian 1995 for a review). Al resistant (*Alr*) and -sensitive (*Als*) mutants have been identified in *Arabidopsis* (Larsen et al. 1997 and references therein) with the *Alr* mutations being semidominant and composed of two independent loci. Out of nine *Als* mutants isolated, eight belonged to different loci, demonstrating the genetic complexity of Al sensitivity in *Arabidopsis*. These mutants were produced by chemical mutagenesis using ethyl methanesulfonate as were the *tir3* mutants published by Ruegger et al. (1997) from the same ecotype (Columbia 1). This opens the possibility to test whether the *Als* mutants (Larsen et al. 1997) that express an increased sensitivity to Al compared to wild type show alterations in NPA binding and the sensitivity of auxin transport to NPA similar to the *tir3* mutant (Ruegger et al. 1997).

Recently, transgenic plants with elevated Al tolerance could be produced in economically important crop species (tobacco and papaya) by overexpression of a citrate synthase (CS) gene from *Pseudomonas aeruginosa* (de la Fuente et al. 1997). The coding sequence of the *P. aeruginosa CS* gene was fused to *35S* promoter from the cauliflower mosaic virus and a *nos 3' terminator* and transfected into the target plant using the Ti-plasmid from *Agrobacterium*. The presence of transgene was confirmed by Southern blot. The transgenic plants synthesized five to six times more citrate than control plants and exhibited a ten-fold higher resistance to Al than the controls. This first approach to produce Al tolerant plants using biotechnological tools was achieved by directly manipulating the expression of a biochemical target for Al.

In order to improve the Al tolerance of growth an alternative approach could be designed the could be based on either manipulating the Al-sensitivity of auxin transport or changing the molecular properties of microtubules such that their response to auxin is altered, and to test the performance of root growth in acidic soils containing Al. Such alterations possibly confer cross-protection to other noxious environmental conditions that affect root growth in a way similar to Al, with outer cortical cells (see Figs. 7 and 8 of Müller et al. 1998).

5.7
Conclusions

From this chapter, the following molecular targets can be identified that allow improvement of Al sensitivity of root growth in crop plants:

1. The number of auxin-efflux carriers (operationally defined as NPA-binding sites) seems to be under genetic control as shown by the *tir3* mutant of *Arabidopsis*, and these genotypic differences seem to affect Al sensitivity.

2. Genotypic differences of endogenous auxin, gibberellin and ethylene content in the root, together with differences of the respective concentration gradients along the root apex, may also determine a genotype to be sensitive or insensitive to Al.

3. Changing the molecular properties of the microtubular cytoskeleton such that the response to the altered hormone content (i.e. disintegration of cortical microtubules) as a consequence of the Al effect on auxin transport could also be used as an approach to manipulate Al sensitivity.

Acknowledgements. We sincerely thank Prof. P. Schopfer, University of Freiburg, for introducing us to the field of cytoskeleton, Prof. Volkmann and Dr. Frantisek Baluška, University of Bonn, and Dr. Hartwig Lüthen, University of Hamburg for carrying out the experiments, the Deutsche Forschungsgemeinschaft (DFG) for financial assistance to WJH, the Deutsche Akademische Austauschdienst (DAAD), Bonn, for an Indo-German postdoctoral fellowship to MS, Dr. C. Weigle (TiHo) Dr. Sanjay Mishra (Institute of Botany) and Dr. G. Grunewaldt (Institute of Plant Pathology and Plant Protection, University of Hannover) for excellent technical assistance. This research was supported in part by a Grant-in-Aid for Encouragement of Young Scientists No.09760042 (1997) and a Grant-in-Aid for Scientific Research No.11660045 (1999) from the Ministry of Education, Science and Culture of Japan.

References

Alfano F, Russell A, Gambardella R (1993) The actin cytoskeleton of the liverwort *Riccia fluitans*: effects of cytochalasin B and aluminium ions on rhizoid tip growth. J Plant Physiol 142:569-574

Archambault DJ, Zhang G, Taylor GJ (1996) Accumulation of Al in root mucilage of an Al-resistant and an Al-sensitive cultivar of wheat. Plant Physiol 112:1471-1478

Baluška F, Hasenstein KH (1997) Root cytoskeleton: its role in perception of and response to gravity. Planta Suppl 203: S69-S78

Baluška F, Volkmann D, Barlow PW (1996) Specialized zones of development in roots: view from the cellular level. Plant Physiol 112: 3-4

Baluška F, Vitha S, Barlow PW, Volkmann D (1997) Rearrangements of F-actin arrays in growing cells of intact maize root apex tissues: a major developmental switch occurs in the postmitotic transition region. Eur J Cell Biol 72: 113-121

Barlow PW, Parker JS (1996) Microtubular cytoskeleton and root morphogenesis. Plant Soil 187: 23-36

Baskin TI, Wilson JE (1997) Inhibitors of protein kinases and phosphatases alter root morphology and disorganize cortical microtubules. Plant Physiol 113: 493-502

Bernasconi P, Patel BC, Reagen JD et al. (1996) The N-1-naphthylphthalamic acid-binding protein is an integral membrane protein. Plant Physiol 111: 427-432

Blamey FPC, Edmeades DC, Wheeler DM (1990) Role of cation-exchange capacity in differential aluminum tolerance of *Lotus* species. J Plant Nutr 13: 728-744

Blancaflor EB, Hasenstein KH (1995) Time course and auxin sensitivity of cortical microtubule reorientation in maize roots. Protoplasma 185: 72-82

Blancaflor EB, Jones DL, Gilroy S (1998) Alterations in the cytoskeleton accompany aluminum induced growth inhibition and morphological changes in primary roots of maize. Plant Physiol 118: 159-172

Bolan NS, Hedley MJ, White RE (1991) Processes of soil acidification during nitrogen cycling with emphasis on legume-based pasture. Plant Soil 134: 53-63

Cox DN, Muday GK (1994) NPA binding activity is peripheral to the plasma membrane and is associated with the cytoskeleton Plant Cell 6: 1941-1953

Cyr R (1994) Microtubules in plant morphogenesis: role of the cortical array. Annu Rev Cell Biol 10: 153-180

De la Fuente JM, Ramirez-Rodriguez V, Carbera-Ponce JL, Herrera-Estrella L (1997) Aluminum tolerance in transgenic plants by alteration of citrate synthesis. Science 276: 1566-1568

Delhaize E, Ryan PR (1995) Aluminum toxicity and tolerance in plants. Plant Physiol 107: 315-321

Delhaize E, Craig S, Beaton CD, Benner RJ, Jagadish VC, Randall PJ (1993a) Aluminum tolerance in wheat (*Triticum aestivum* L.) I. Uptake and distribution of aluminum in root apices. Plant Physiol 103: 685-693

Delhaize E, Ryan PR, Randall PJ (1993b) Aluminum tolerance in wheat (*Triticum aestivum* L.) II. Aluminum-stimulated excretion of malic acid from root apices. Plant Physiol 103: 695-702

Estelle M (1998) Polar auxin transport: new support for an old model. Plant Cell 10: 1775-1778

Evans ML, Ishikawa H (1997) Cellular specificity of the gravitropic motor response in roots. Planta 203: 115-122

Fisher DD, Cyr RJ (1998) Extending microtubule/microfibril paradigm. Cellulose synthesis is required for normal cortical microtubule alignment in elongating cells. Plant Physiol 116: 1043-1051

Foy CD (1988) Plant adaptation to acid, aluminium-toxic soils. Commun Soil Sci Plant Anal 19: 959-987

Giddings TH, Staehelin LA (1991) Microtubule-mediated control of microfibril deposition: a re-examination of the hypothesis. In: Lloyd CW (ed) The cytoskeletal basis of plant growth and form. Academic Press, London, pp 85-99

Grabski S, Schindler M (1995) Aluminum induces rigor within the actin network of soybean cells. Plant Physiol 108: 897-901

Grabski S, Schindler M (1996) Auxins and cytokinins as antipodal modulators of elasticity within the actin network of plant cells. Plant Physiol 110: 965-970

Grabski S, Arnoys E, Busch B, Schindler M (1998) Regulation of actin tension in plant cells by kinases and phosphatases. Plant Physiol 116: 279-290

Gunning BES, Hardham AR (1982) Microtubules. Annu Rev Plant Physiol 33: 651-698

Hasenstein KH, Evans ML (1988) Effects of cations on hormone transport in primary roots of *Zea mays*. Plant Physiol 86: 890-894

Hasenstein KH, Blancaflor EB, Lee JS (1999) The microtuble cytoskeleton does not integrate auxin transport and gravitropism in maize roots. Physiol Plant 105: 729-738

Haug A (1984) Molecular aspects of aluminum toxicity. Crit Rev Plant Sci 1: 345-373

Horst WJ (1995) The role of the apoplast in aluminum toxicity and resistance of higher plants: a review. Z Pflanzenernähr Bodenkd 158: 419-428

Horst WJ, Wagner A, Marschner H (1982) Mucilage protects root meristems from aluminium injury. Z Pflanzenphysiol 109: 95-103

Horst WJ, Püschel A-K, Schmohl N (1997) Induction of callose formation is a sensitive marker for genotypic aluminium sensitivity in maize. Plant Soil 192: 23-30

Horst WJ, Schmohl N, Kollmeier M, Baluška F, Sivaguru M (1999) Does aluminum inhibit root growth through the interaction with the cell wall-plasma membrane-cytoskeleton continuum? Plant Soil 215: 163-174

Jones DL, Shaff JE, Kochian LV (1995) Role of calcium and other ions in directing root hair tip growth in *Limnobium stoloniferum*. I. Inhibition of tip growth by aluminum. Planta 197: 672-680

Jones DL, Kochian LV, Gilroy S (1998a) Aluminum induces a decrease in cytosolic calcium concentration in BY-2 tobacco cell cultures. Plant Physiol 116: 81-89

Jones DL, Gilroy S, Larsen PB, Howell SH, Kochian LV (1998b) Effect of aluminum on cytoplasmic Ca^{2+} homeostasis in root hairs of *Arabidopsis thaliana* (L.). Planta 206: 378-387

Kaufmann PB, Wu LL, Brock TG, Kim D (1995) Hormones and the orientation of growth. In: Davies PJ (ed) Plant hormones – physiology, biochemistry and molecular biology. Kluwer, Dordrecht, pp 547-578

Kauss H (1996) Callose synthesis. In: Smallwood M, Knox JP, Bowles DJ (eds) Membranes: special-ised functions in plants. Bios, London, pp 77-92

Kerk N, Feldman L (1994) The quiescent center in roots of maize: initiation, maintenance and role in organization of the root apical meristem. Protoplasma 183: 100-106

Kinraide TB (1993) Aluminum enhancement of plant growth in acid rooting media. A case of recipro-cal alleviation of toxicity by two toxic cations. Physiol Plant 88: 619-625

Kinraide TB, Ryan PR, Kochian LV (1992) Interaction effects of Al^{3+}, H^+, and other cations on root elongation considered in terms of cell-surface electrical potential. Plant Physiol 99: 1461-1468Kochian LV (1995) Cellular mechanisms of aluminum toxicity and resistance in plants. Annu Rev Plant Physiol Plant Mol Biol 46: 237-260

Kollmeier M, Felle HH, Horst WJ (2000) Genotypical differences in Al resistance of Zea mays (L.) are expressed in the distal part of the transition zone. Is reduced basipetal flow involved in inhibition of root elongation by Al? Plant Physiol (in press)

Lagrimini LM, Joly RJ, Dunlap JR, Liu TTY (1997) The consequence of peroxidase overexpression in transgenic plants on root growth and development. Plant Mol Biol 33: 887-895

Larsen PB, Stenzler LM, Tai CY, Degenhardt J, Howell SH, Kochian LV (1997) Molecular and physiological analysis of Arabidopsis mutants exhibiting altered sensitivities to aluminum. Plant Soil 192: 3-7

Lazof DB, Goldsmith JG, Rufty TW Linton RW (1994) Rapid uptake of aluminum into cells of intact soybean root tips. A microanalytical study using secondary ion mass spectroscopy. Plant Physiol 106: 1107-1114

Lindberg S, Strid H (1997) Aluminium induces rapid changes in cytosolic pH and free calcium and potassium concentrations in root protoplasts of wheat (Triticum aestivum). Physiol Plant 99: 405-414

Lloyd CW (1991) The cytoskeletal basis of plant growth and form. Academic Press, London

Llugany M, Massot N, Wissemeier AH, Poschenrieder C, Horst WJ, Barcelo J (1994) Aluminum tolerance of maize cultivars as assessed by callose production and root elongation. Z Pflanzen-ernähr Bodenkd 157: 447-451

Ma JF, Zhen SJ, Matsumoto H (1998) High aluminum resistance in buckwheat II. Oxalic acid detoxi-fies aluminum internally. Plant Physiol 117: 753-759

MacDonald TL, Humphreys WG, Martin RB (1987) Promotion of tubulin assembly by aluminum ion in vitro. Science 236: 183-186

Marienfeld S, Schmohl N, Klein M, Schröder WH, Kuhn AJ, Horst WJ (2000) Localisation of alumi-num in root tips of Zea mays and Vicia faba. J Plant Physiol (in press)

Matsumoto H (2000) Cell biology of Al tolerance and toxicity in higher plants. Int Rev Cytol (in press)

Miller D, Hable W, Gottwald J, Ellard-Ivey M Demura T, Lomax T, Carpita N (1997) Connections: the hard wiring of the plant cell for perception, signaling, and response. Plant Cell 9: 2105-2117

Mizuno K (1992) Induction of cold stability of microtubules in cultured tobacco cells. Plant Physiol 100: 740-748

Müller A, Guan C, Gälweiler L, Tänzler P, Huijser P, Marchant A, Parry G, Bennett M, Wisman E, Palme K (1998) AtPIN2 defines a locus of Arabidopsis for root gravitropism control. EMBO J 17: 6903-6911

Nouri PA, Reddy GB (1995) Influence of acid rain and ozone on heavy metals under loblolly pine trees: a field study. Plant Soil 171: 59-62

Papernik LA, Kochian LV (1997) Possible involvement of Al-induced electrical signals in Al tolerance in wheat. Plant Physiol 115: 657-667

Pellet DM, Grunes DL, Kochian LV (1995) Organic acid exudation as an aluminum-tolerance mecha-nism in maize (Zea mays L.). Planta 196: 788-795

Pettersson S, Strid H (1989) Effects of aluminium on growth and kinetics of K^+ [^{86}Rb] uptake in two cultivars of wheat (Triticum aestivum) with different sensitivity to aluminium. Physiol Plant 76: 255-261

Reid RJ, Tester MA, Smith FA (1995) Calcium/aluminum interactions in the cell wall and plasma membrane of Chara. Planta 195: 362-368

Rengel Z (1996) Uptake of aluminium by plant cells. New Phytol 134: 389-406

Ruegger M, Dewey E, Hobbie L (1997) Reduced naphthylphthalamic acid binding in the *tir3* mutant of *Arabidopsis* is associated with a regulation in polar auxin transport and diverse morphological defects. Plant Cell 9: 745-757

Ryan PR, DiTomaso JM, Kochian LV (1993) Aluminium toxicity in roots: an investigation of spatial sensitivity and the role of the root cap. J Exp Bot 44: 437-446

Sanchez PA, Logan TJ (1992) Myths and science about the chemistry and fertility of soils in the tropics. In: Lal RS, Sanchez PA (eds) Myths and science of soils of the tropics. SSSA special publication No 29, Soil Sci Soc Am and Am Soc Agric, Madison, WI, pp 35-36

Sasaki M, Yamamoto Y, Matsumoto H (1997) Aluminum inhibits growth and stability of cortical microtubules in wheat (*Triticum aestivum*) roots. Soil Sci. Plant Nutr 43: 469-472

Schmidt R, Bohm K, Vater W, Unger E (1991) Aluminum-induced osteomalacia and encelopathy: an aberration of the tubulin assembly into microtubules by Al^{3+}. Prog Histochem Cytochem 23: 355-364

Shibaoka H (1994) Plant hormone-induced changes in the orientation of cortical microtubules: alterations in the cross-linking between microtubules and the plasma membrane. Annu Rev Plant Physiol Mol Biol 45: 527-544

Simmonds DH (1992) Plant cell wall removal: cause for microtubule instability and division abnormalities in protoplast cultures? Physiol Plant 85: 387-390

Sivaguru M, Horst WJ (1998) The distal part of the transition zone is the most aluminium-sensitive apical root zone of *Zea mays* L. Plant Physiol 116: 155-163

Sivaguru M, Baluška F, Volkmann D, Felle H, Horst WJ (1999a) Impacts of aluminum on the maize root cytoskeleton. Short-term effects on the distal part of the transition zone. Plant Physiol 119: 1073-1082

Sivaguru M, Yamamoto Y, Matsumoto H (1999b) Differential impacts of aluminum on microtubules depends on the growth phase in suspension cultured tobacco cells. Physiol Plant 107: 110-119

Takabatake R, Shimmen T (1997) Inhibition of electrogenesis by aluminum in characean cells. Plant Cell Physiol 38: 1264-1271

Taylor GJ (1995) Overcoming barriers to understanding the cellular basis of aluminum resistance. Plant Soil 171: 89-103

Tice KR, Parker DR, DeMason A (1992) Operationally defined apoplastic and symplastic aluminum fractions in root tips of aluminum-intoxicated wheat. Plant Physiol 100: 309-318

Vázquez MD, Poschenrieder C, Corrales I, Barceló J (1999) Changes in apoplastic aluminum during the initial growth response to aluminum by roots of a tolerant maize variety. Plant Physiol 119: 435-444

von Uexküll HR, Mutert E (1995) Global extent, development and economic impact of acid soils. Plant Soil 171: 1-15

Williamson RE (1991) Orientation of cortical microtubules in interphase plant cells. Int Rev Cytol 129: 135-206

Wissemeier AH, Klotz F, Horst WJ (1987) Aluminium induced callose synthesis in roots of soybean (*Glycine max* L.). J Plant Physiol 129: 487-492

Wissemeier AH, Horst WJ (1995) Effect of calcium supply on aluminium-induced callose formation, its distribution and persistence in roots of soybean (*Glycine max* (L.) Merr.). Journal of Plant Physiology 145: 470-476.

Wyatt SE, Carpita NC (1993) The plant cytoskeleton-cell wall continuum. Trends Cell Biol 3: 413-417

Young LM, Evans ML (1994) Calcium-dependent asymmetric movement of ^{3}H-indole-3 acetic acid across gravistimulated isolated root caps of maize. Plant Growth Regulation 14: 235-242

Zhang G, Taylor GJ (1989) Kinetics of aluminum uptake by excised roots of aluminum-tolerant and aluminum-sensitive cultivars of *Triticum aestivum* L. Plant Physiol 91: 1094-1099

Zhang G, Taylor GJ (1990) Kinetics of aluminum uptake in *Triticum aestivum* L. Identity of the linear phase of aluminum uptake by excised roots of aluminum-tolerant and aluminum-sensitive cultivars. Plant Physiol 94: 577-584

Zhang G, Hoddinott T, Taylor GJ (1994) Characterization of 1-3-β-glucan (callose) synthesis in roots of *Triticum aestivum* in response to aluminum toxicity. J Plant Physiol 144: 229-234

Zhang,G.; Slaski,J.J.; Archambault,D.J.; Taylor,G.J. (1997) Alteration of plasma membrane lipids in aluminium-resistant and aluminium- sensitive wheat genotypes in response to aluminium stress. Physiol .Plant 99, 302-308

Zhang W, Rengel Z, Kuo J (1998) Determination of intracellular Ca^{2+} in cells of intact wheat roots: loading of acetoxymenthyl ester of Fluo-3 under low temperature. Plant J 15: 147-151

6 Control of the Response to Low Temperatures

Peter Nick
Institut für Biologie II, Schänzlestr. 1, D-79104 Freiburg, Germany

6.1
Summary

Low temperatures limit the productivity of a number of cultivated plant species especially in temperate regions. Microtubules seem to play a dual role in this response –on the one hand, they depolymerize in response to low, non-freezing temperatures and this is correlated with corresponding changes of cell growth; on the other hand, there is evidence for a role of microtubule depolymerization in the sensing process itself, an aspect that is relevant to cryopreservation of tissue and seeds in gene banks. The cold stability of microtubules varies between species and can be regulated by hormones such as abscisic acid and by physiological responses such as cold acclimation. The tubulin domains that are responsible for the microtubular response to low temperature can be inferred from a comparison of tubulin isotypes with inherent differences in cold stability. The mediating signal chains seem to involve calcium and calmodulin. This chapter concludes with an outlook on biotechnological strategies to manipulate the cold response of microtubules and thus the cold resistance of plant organs or developmental processes.

6.2
The impact of low temperature in agriculture

In temperate regions, temperature poses major constraints to crop yield. Attempts to increase photosynthetic rates by conventional breeding programs, although pursued over a long period, have failed so far, indicating that evolution has already reached the optimal situation (Evans 1975). However, the period of the year where these photosynthetic rates are actually reached is reduced by low temperatures. During the spring season, although the quantity of light would allow for high photosynthetic rates, the retarded development of leaves and thus reduced leaf area limits productivity (Watson 1952). Interestingly, the cold sensitivity of photosynthesis is much less pronounced than that of growth. In other words: in temperate regions it is the cold sensitivity of growth that has to be regarded as the limiting factor for productivity (Monteith and Elston 1971). This conclusion is supported by the finding that in cool climates the production of biomass is sink limited rather than source-limited (Warren-Wilson 1966). The main target seems to be the root – it is the temperature of the root that confines the speed of shoot development (Atkin et al. 1973).

Sensitivity to low temperature is an agricultural issue that is not confined to the temperate regions. Many tropical and subtropical plants suffer severely when they are exposed to cool temperatures that are still far above the freezing point. This poses extreme problems when fruits have to be harvested and cooled for transport and processing, because these fruits rot rapidly as soon as they return to warmer temperatures. This phenomenon has been known for a long time and was termed originally Erkältung (chilling damage) by Molisch (1897) to distinguish it from damages that are caused by actually freezing the tissue. In some cases, even higher temperatures can cause irreversible harm, if they hit a sensitive period of development. As an example, the fertility of rice is extremely and irreversibly reduced when temperatures fall below 18°C during flower development with drastic economic consequences. Estimates of the Japanese Ministry of Agriculture, Forestry and Fishery on the yield losses in rice during the cool summer of 1993 are in the range of 25% (corresponding to half a billion ECU).

The extent of cold sensitivity can thus vary considerably between different species and even between different cultivars of the same crop. The term chilling resistance is used for plants that can cope with cool, but non-freezing temperatures (below 10 °C), whereas freezing resistant plants such as rye or winter wheat can survive temperatures even below zero (Lyons 1973). It should be kept in mind that the degree of cold sensitivity can change depending on development and environmental conditions. For instance, the freezing tolerance of many species can be increased in consequence of preceding cool, but non-freezing periods and this so-called cold acclimation or cold hardening determines the survival during the winter (Stair et al. 1998).

The problem of cold tolerance is not only important from a biotechnological point of view but poses an interesting scientific issue as well, because the response to low temperature involves the transformation of a physical signal into a cascade of biochemical signalling events. This process is still far from being understood. For this reason, the physiological responses to low temperatures will be described in the following section.

6.3
Cellular effects of low temperatures

As pointed out above, it is important to make a clear distinction between freezing injury (temperatures below 0 °C) and chilling injury (temperatures ranging between 0 and 10 °C).

The cellular consequences of freezing injury are more obvious and better understood. In the case of rapid freezing, ice crystals form within the cells and disrupt internal and external membranes, which will kill the cell instantaneously (Burke et al. 1976). As long as the freezing occurs at a slow pace and does not exceed a certain limit, ice will form outside the cell and will remain on the surface of cell walls, in vessel elements and on the external surface. This does not kill the cells as long as the crystals do not penetrate the plasma membrane. However, the plant will dry out in the long term because the access of water to the roots is impaired (Mazur 1963).

Plants that can reduce dehydration because they can minimize transpiration can cope better with freezing temperatures. Therefore subarctic or subalpine forests are dominated by xeromorphic species (such as the conifers). A second strategy against freezing injury seems to be the expression of cold-regulated genes that encode unique hydrophilic polypeptides (Hughes and Dunn 1996). Some of these polypeptides reveal similarities with antifreezing proteins from antarctic fishes (Kurkela and Franck 1990), indicating that they are shifting the phase transition towards the solid state in membranes and cytoplasm. Recently, one of these proteins, COR15a, has been suggested to extend the lamellar phase of chloroplasts to lower temperatures (Steponkus et al. 1998).

The actual reason for chilling injury is less evident and has not received the same degree of attention as freezing injury. An apparently trivial consequence of low temperature is a reduced rate of biochemical processes. This should result in a reduced metabolic activity, but cannot account for irreversible damages. However, the temperature dependence can vary considerably between different enzymes (Guy 1990): rice tonoplast H^+-ATPase is blocked at 10 °C, potato phosphofructo kinase or maize starch synthetase already at 12 °C, whereas maize PEP carboxylase is still active at 4 °C. When the temperature decreases below 10 °C, it is mainly the enzymes of the respiratory chain that are affected, possibly because they are membrane-bound, whereas some of the glycolytic enzymes are still active. This will result in metabolic imbalance and the accumulation of ethanol and acetaldehyde (Lyons 1973).

Membrane-bound enzymes are more susceptible to chilling injury as compared to soluble enzymes. The reason has to be sought in the fluidity of the membranes (Lyons 1973), and membrane fluidity is tightly coupled with the presence of unsaturated fatty acids. A membrane composed exclusively of a fully saturated lipid species should exhibit phase separation at around 30 °C, whereas the introduction of one *cis*-double bound in the centre of the molecule would decrease the phase transition temperature to 0 °C (Ishizaki-Nishizawa et al. 1996). The degree of lipid saturation can be manipulated by overexpression of desaturases in plants and this has been shown repeatedly to increase chilling sensitivity (Murata et al. 1992; Wolter et al. 1992; Kodama et al. 1994; Ishizaki-Nishizawa et al. 1996).

One of the most pronounced and rapid cellular indications of chilling injury is the cessation of cytoplasmic streaming (Kamiya 1959; Woods et al. 1984; Tucker and Allen 1986) which occurs within a few minutes when temperature falls to 10 °C in chilling-sensitive species such as cucumber or tomato (Sachs 1865), whereas it can proceed down to 0 °C in chilling-resistant plants (Lyons 1973). When the period of chilling exceeds a few hours, cytoplasmic streaming fails to recover. Kinetic studies in subtropical species such as maize, lima bean and cotton (Lyons 1973) showed developmental differences with a high sensitivity observed in those periods, where the rate of cell growth is high. Additionally, a high chilling sensitivity is characteristic for the morphogenetic response to light such as axis formation in lower plants (Haupt 1958) or phototropism in maize (Nick and Schäfer 1991).

The cellular effects of low temperature can be classified according to their sensitivity (Fig. 6.1): those events that are involved in the maintenance or control of cell growth seem to be among the most sensitive processes in chilling sensitive species that are affected already at temperatures above 12 °C. When the temperature is lowered, membrane fluidity and the activity of membrane-bound enzymes (respiration chain, photosynthesis) are the processes that are impaired next, whereas the block of glycolysis and the damage to membranes by ice crystals are processes that might become limiting only chilling-resistant species. Cell growth is intimately linked to the microtubular cytoskeleton, as pointed out in Chapter 1, this book. It is therefore worth considering whether microtubules are a target of low temperature.

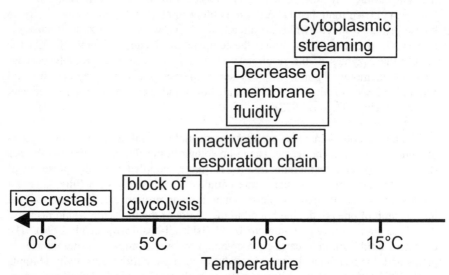

Fig.6.1. Classification of different cellular and metabolic events according to their cold sensitivity. Cytoplasmic streaming and changes in membrane fluidity are among the most sensitive processes, and might be the limiting steps in the cold response of chilling-sensitive species, whereas glycolysis and membrane ruptures produced by ice crystals occur at much lower temperatures and are relevant for freezing tolerance rather than chilling tolerance.

6.4
Microtubules as target of low temperatures

Microtubules of both plants and animals disassemble in response to low temperature, but the degree of cold sensitivity differs depending on the type of organism. Whereas mammalian microtubules disassemble when the temperature falls below +20 °C, the microtubules from poikilothermic animals of temperate regions remain intact below this temperature (Modig et al. 1994). In plants, cold stability of microtubules seems to be more pronounced as compared to animals (Juniper and Lawton 1979), reflecting the higher developmental plasticity of plants. However, the critical temperature that can induce microtubule disassembly varies considerably between different species (Jian et al. 1989, Chu et al. 1992, Pihakaski-Maunsbach and Puhakainen 1995): in chilling-sensitive plants such as maize, tomato or cucumber, microtubules disassemble already at temperatures above 0-4 °C, whereas they can withstand 0 °C in moderately resistant plants such as spinach and beet. In cold-resistant species such as winter wheat or winter rye even temperatures down to −5 °C are not sufficient to eliminate microtubules. Even within a given species, the cold sensitivity of microtubules can vary considerably, and the degree of microtubular cold hardiness can be related to the resistance of growth to low temperature (Fig. 6.2): whereas microtubules in the root elongation zone of the Mexican wheat cultivar Siete Cerro disassemble already when exposed to 1 hour at 4 °C, the microtubules of the Chinese winter wheat cultivar Jing Nong 934 are still present after 1 h at 0 °C and microtubules of the Russian winter wheat cultivar Mironovska 808 survive even at −5 °C (Fig. 6.2a) related to corresponding responses of root elongation (Fig. 6.2b).

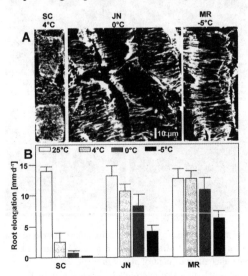

Fig.6.2. Differences in the cold sensitivity of microtubules (**a**) and root elongation (**b**) in the wheat cultivars Siete Cerro (**SC**, sensitive), Jing Nong (**JN**, chilling-resistant) and Mironovska 808 (**MR**, freezing resistant). Microtubules in the outer cortex of the root elongation zone are shown in **a**. The cold shock was of 2-h duration.

The close connection between microtubular cold sensitivity and the chilling sensitivity of cell growth is supported by the observation that abscisic acid, a hormonal inducer of cold hardiness (Holubowicz and Boe 1969; Irving 1969; Rikin et al. 1975; Rikin and Richmond 1976), can stabilize cortical microtubules against low temperature (Sakiyama and Shibaoka 1990). Conversely, the destabilization of microtubules by assembly blockers such as colchicine or podophyllotoxin increased the chilling sensitivity of cotton seedlings, and this effect could be reversed by addition of abscisic acid (Rikin et al. 1980). Gibberellin, a hormone that has been shown in several species to decrease cold hardiness (Rikin et al. 1975; Irving and Lanphear 1986), renders cortical microtubules cold-labile (Akashi and Shibaoka 1987).

It is possible to increase the cold resistance of an otherwise chilling-sensitive species by precultivation at moderately cool temperature (so-called cold hardening). The genes that are activated during cold hardening are partially identical to those that respond to abscisic acid (reviewed in Hughes and Dunn 1996), and the tissue content of abscisic acid increases in response to cold hardening (Lalk and Dörffling 1985; Lång et al. 1994). On the other hand, ABA-insensitive mutants are capable of cold hardening to a certain extent (Gilmour and Thomashow 1991), indicating the coexistence of ABA-independent pathways in addition to the ABA dependent pathway triggering cold hardening.

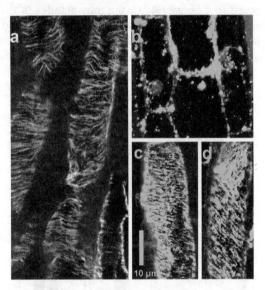

Fig.6.3a-d. Response of microtubules in outer cortical cells of the elongation zone in roots of the freezing-tolerant winter wheat cultivar Mironovska 808 to a cold shock of –7 °C lasting for 2 h. **a** Microtubules in unadapted unchallenged control plants cultivated at 25°C. **b** microtubules in unadapted control plants after the cold shock; note that microtubules are completely disassembled. **c** Microtubules of cold-adapted plants (0 °C, 12 h); after the cold shock they are transverse and disperse. **d** Microtubules of ABA-pretreated plants (10 μM, 12 h); after the cold shock they are oblique and bundled.

Cold hardening can be detected on the level of microtubules as well. Microtubules of cold-acclimated spinach mesophyll cells coped better with the consequences of a freeze-thaw cycle (Bartolo and Carter 1991a). Although abscisic acid can increase the cold resistance of microtubules (Sakiyama and Shibaoka 1990), this seems not to be the only trigger. Striking differences can be observed in roots of winter wheat when the microtubular response to abscisic acid is compared to the effect of cold hardening (Fig. 6.3): whereas the increased microtubular cold resistance induced by abscisic acid is accompanied by a reorientation into longitudinal arrays and a conspicuous bundling of microtubules, the microtubules maintain a transverse array of fine strands when the cold resistance is induced by precultivation at moderately cool temperatures. This indicates that microtubular cold hardening is induced by an ABA-independent pathway.

The findings discussed so far support a model where the sensitivity of microtubules to low temperature limits the chilling (not freezing!) resistance of growth. This is expected to be of ecological relevance in periods of increased chilling sensitivity (for instance during seedling growth) or in chilling-sensitive species (such as rice, cucumber or cotton). Treatments that increase the cold resistance of microtubules can, under these conditions, improve the performance of the whole plant under low temperature.

6.5
Microtubules as modulators of cold sensing

Microtubules are not only target of cold stress, they seem, in addition, to participate in cold sensing itself, triggering a chain of events that culminates in increased cold hardiness. This function during the sensory process is mainly illustrated by two lines of evidence.

1.   Suppression of microtubule depolymerization by taxol results in reduced freezing tolerance and the failure to induce cold hardiness (Kerr and Carter 1990; Bartolo and Carter 1991b). This indicates that microtubules have to disassemble to a certain degree in order to trigger cold acclimation. This partial disassembly, however, is a transient, early event. Upon prolonged cold exposure, the microtubular cytoskeleton recovers and exhibits, in the long term, a reduced disassembly in response to low temperature (Pihakaski-Maunsbach and Puhakainen 1995). This transient nature of the disassembly response is characteristic for events that are related to signal sensing.
2.   Cold shock causes a rapid and transient rise of intracellular calcium concentration (Knight et al. 1991). This could be elegantly demonstrated elegantly using transgenic plants expressing aequorin and measuring bioluminescence in response to cold shock. Pharmacological data (Monroy et al. 1993) demonstrate that this calcium peak is not only a byproduct of the cold response, but necessary to trigger cold acclimation. The formation of this peak involves calcium channels as well as calmodulin. These data suggest a central role of cold-sensitive calcium channels for the induction of cold hardening. The activity of calcium channels, on the other

hand, can be modulated via a manipulation of microtubules. If microtubules are disassembled by drugs such as oryzalin or colchicin, this leads to a six- to tenfold fold increase in the activity of voltage-dependent calcium channels (Ding and Pickard 1993; Thion et al. 1996), and, specifically, to a conspicuous enhancement of cold-induced calcium fluxes (Mazars et al. 1997).

These studies suggest that microtubules function as modulators of cold sensing (Fig. 6.4): stable microtubules limit the activity of cold-induced voltage-dependent calcium channels. When microtubules disassemble, however, this causes a release of this constraint and the activity of the channels becomes elevated accelerating the signal chain that culminates in increased cold hardiness. In the long term, the increased cold hardiness is likely to feed back upon microtubular stability (see previous section) and, indirectly, on the capacity of the cold-inducible channels. As discussed in the following section, disassembly of microtubules, in turn, can be induced by binding of calcium/calmodulin to microtubules. Such a feedback loop could establish a regulatory network that allows for very fine and precise tuning of the cellular response to low temperature. To understand this feedback loop in more detail, it is necessary to consider some of its molecular components.

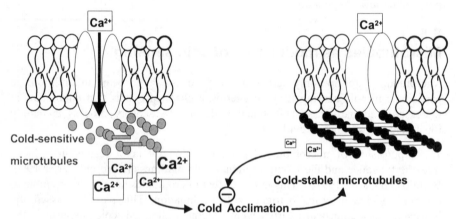

Fig.6.4. Model for the role of microtubules in the feedback circle controlling cold hardening. In unadapted cells, cold stress induces the opening of specific calcium channels, resulting in a flux of calcium into the cell triggering disassembly of microtubules. The disassembly of microtubules enhances the activity of these calcium channels. The rise of intracellular calcium concentration induces a signal chain, resulting in cold acclimation leading to a production of cold-stable microtubules that confine the activity of cold-sensitive calcium channels. This contributes to a reduction of intracellular calcium levels, which, in turn, will weaken the induction of cold acclimation.

6.6
Molecular aspects of the microtubular cold response

As mentioned above, the cold sensitivity of microtubules varies between different organisms. Moreover, it can be regulated during development and in response to signals as, for instance, during the cold-hardening response. Although a comprehensive picture has not emerged so far, some of the molecular components involved in cold-induced microtubule disassembly have been identified.

1. One of the factors involved in the cold response of microtubules seems to be calcium acting via the calcium/calmodulin pathway. Immunocytochemical data suggest a binding of calmodulin to cortical microtubules depending on the concentration of calcium (Fisher and Cyr 1993). The dynamics of microtubules could be regulated via a calmodulin-sensitive interaction between microtubules and microtubule-associated proteins such as the bundling protein EF-1α (Durso and Cyr 1994). Alternatively, the calcium/calmodulin pathway could directly couple to the C-terminus of tubulin because proteolytic cleavage of the C-terminus renders maize microtubules resistant to both low temperature and calcium (Bokros et al. 1996). If the release of calcium from intracellular pools is blocked by lithium, an inhibitor of polyphosphoinositide turnover (Berridge and Irvine 1984), cold-induced disassembly of mesophyll microtubules was observed to decrease (Bartolo and Carter 1992), consistent with a role for calcium in the microtubular low-temperature response.

2. Cold stability of microtubules can be induced by inhibitors of protein kinases such as staurosporin or 6-dimethylaminopurine (Mizuno 1992) in young tobacco cell cultures, suggesting that cold sensitivity is actively maintained by a pathway involving protein kinases. This cold sensitivity might be related to a high turnover of assembly and disassembly that seems to be characteristic for cortical microtubules in actively growing cells (Wasteneys et al. 1993; Yuan et al. 1994; Himmelspach et al. 1999). In older cells, where the microtubular dynamics becomes reduced, microtubules acquire a natural cold stability (Mizuno 1992), possibly by cross-bridges between cortical cytoskeleton and cell wall probably involving extensins (Akashi et al. 1990).

3. Cold hardening of microtubules might involve expression of differential tubulin isotypes. In rye roots, the pattern of tubulin isotypes has been described to respond fairly rapidly to cold acclimation (Kerr and Carter 1990). In *Arabidopsis*, the expression of several tubulin genes changes in response to cold acclimation. Some isotypes disappear (such as TUB2, TUB3, TUB6 and TUB8), whereas one isotype (TUB9) is upregulated (Chu et al. 1993b). Interestingly, the TUB9 gene product is characterized by a shorter C-terminus (for details refer to Chap. 7, this book). Consistent with this cold-induced expression of new isotypes, an increase of HSP90, a marker for microtubule nucleation (Petrášek et al. 1998) has been reported for *Brassica napus* (Krishna et al. 1995).

These observations suggest a central role for the C-terminal domain of tubulin in the cold-induced disassembly of microtubules. It is conceivable that calcium released from intracellular stores in response to low temperature interacts with the C-terminus of tubulin (Bokros et al. 1996) via either calmodulin (Fisher and Cyr 1993) or modulation of a kinase cascade (Mizuno 1992) that might involve calcium-dependent protein kinases such as the maize CDPK1 (Monroy and Dhindsa 1995; Berberich and Kusano 1997). A further factor that is expected to influence cold sensitivity of microtubules is the dynamics of microtubular turn-over. Dynamic microtubules are more sensitive to low temperature, whereas microtubules that are bundled by microtubule-associated proteins (Durso and Cyr 1994) or stabilized by cross-linking proteins (Akashi et al. 1990) are characterized by cold stability.

The C-terminus of tubulin and modulators of microtubular dynamics could thus be used as targets to manipulate the response of plant cells to low temperature.

6.7
Perspectives

6.7.1
Improved cryopreservation by manipulation of microtubules?

Gene banks have started to use, in addition to seeds, frozen tissue samples in order to conserve of genetic and metabolic diversity. Cryopreservation of plant cells has to face the problem that the often highly vacuolized cells suffer from considerable physical stresses such as high osmolarity or alterations in shape and volume. A second critical step is the survival of frozen cells after thawing. It is possible to improve cryosurvival by manipulation of the cytoskeleton.

Treatment of rye roots (Chu et al. 1993a) or spinach leaves (Bartolo and Carter 1991b) with taxol resulted in a stabilization of both microtubules (Bartolo and Carter 1991b) and, indirectly, actin microfilaments (Chu et al. 1993a). The suppression of cytoskeletal dynamics induced by this drug resulted in a reduced rate of cryosurvival (Bartolo and Carter 1991b). In contrast, if the integrity of the cytoskeleton was artificially reduced using cytochalasins, this was accompanied by an improved preservation in liquid nitrogen (Morisset et al. 1993, 1994). Moreover, additives that are known to improve cryopreservation such as dimethylsulfoxide have been shown to induce a partial disintegration of the cytoskeleton (Morisset et al. 1994). This suggests that a certain degree of cytoskeletal disassembly is beneficial to improve cryosurvival of plant cells.

The cause for this surprising phenomenon might be related to the modulating function of the cytoskeleton in cold sensing and in the activity of cold-inducible calcium channels (Thion et al. 1996; Mazars et al. 1997). By cold-induced microtubule disassembly the signal chain triggered by the cooling process might be stimulated, leading to a more efficient induction of cold hardening that could be responsible for the improved performance of the cells during thawing.

This phenomenon could be utilized to optimize cryopreservation. A simple application might consist of a mild pretreatment prior to freezing with microtubule polymerization blockers such as carbamate or dinitroaniline herbicides or, even simpler and cheaper, with a mild chilling pulse. The resulting partial disassembly of microtubules is expected to induce or at least amplify the signalling chains that are induced during freezing and thawing and that are responsible for cold acclimation.

6.7.2
Improved cold tolerance by designed tubulins?

The target domain for cold-induced signalling on the tubulin protein seems to be the C-terminus (see above). Cleavage of the C-terminus renders maize microtubules resistant to low temperature and calcium (Bokros et al. 1996), and the cold induced TUB9 gene product of *Arabidopsis* is characterized by a C-terminal reduction (Chu et al. 1993b).

The pattern of tubulin isotypes rapidly responds to cold shock (Kerr and Carter 1990; Chu et al. 1993b), indicating that the cold tolerance of microtubules could be manipulated by changing the molecular properties of tubulins. The structure and sequence of tubulins has been well preserved during evolution, indicating that it might prove difficult to manipulate tubulin sequences without affecting vital tubulin functions such as GTP-binding or dimerization (see Chap. 7, this book for details). In contrast, the C-terminus of tubulins has been found to be more variable between different organisms and between different isotypes within a given species. This probably mirrors a possible function of the C-terminus in signal-dependent regulation of tubulin dynamics.

It seems therefore feasible to produce designed tubulins where the C-terminus is either truncated (nonsense mutations in the 3'-end of the tubulin gene) or, in a more subtle version, where individual amino acids in the C-terminus are exchanged (missense mutations in the 3'-end of the tubulin gene). These designed tubulin variants could then be expressed, for instance, using the system described in Chap. 7 for transgenic plants and be tested for their effect on cold sensitivity of microtubules. Recently, a mutated tubulin from goosegrass (*Eleusine indica*) has been expressed successfully in a maize suspension culture and the molecular properties of the goosegrass tubulin (resistance to the herbicide oryzalin) were transferred to the microtubules of the host cells (Anthony et al. 1998). However, these experiments had shown that an imbalance between α- and β-tubulin levels is not tolerated by the plant cell (Anthony and Hussey 1998). This means that along with the 3'-modified or truncated gene for α-tubulin a full-length gene for β-tubulin has to be expressed in tandem under control of the same promotor or at least under a promotor of comparable strength to reach balanced expression of both tubulin types.

6.7.3
How to avoid chemical sprouting inhibitors in potatoes

The world production of potatoes ranges around 275 Mio tons per year according to estimates of the International Potato Institute in Lima, Peru. About 60% of the harvest is either exported, stored or processed, posing the problem that sprouting of these potatoes has to be suppressed by inhibitors such as phenylurethane, propham or chlorpropham. The usual doses range around 5-10 g per 100 kg of potatoes. Whereas propham has been widely used in Eastern Europe, Northern Africa and Asian countries, it has been banned in Western Europe for ecological concerns and replaced by chlorpropham, although it remains unclear to what extent chlorpropham represents an improvement. Alternatively, potatoes are stocked at low temperature (1-3 °C) to suppress or at least delay sprouting. However, this is accompanied by undesired sweetening and darkening of potatoes via the nonenzymatic Maillard reaction.

The biological effect of the sprouting inhibitors phenylurethane, propham and chlorpropham is based upon an inhibition of mitosis (Ivens and Blackman 1950) and cell growth (Templeman and Sexton 1945) and was shown to be caused by a drug-induced depolymerization of microtubules (Shibaoka and Hogetsu 1977). Similarly, the inhibition of sprouting by low temperature is expected to involve depolymerization of microtubules as well.

Since the cold sensitivity of microtubules seems to reside in the C-terminal domain (see above), it might be possible to create microtubules with increased cold sensitivity by introducing into the C-terminus of the α-tubulin isotype expressed in tubers and germinating sprouts peptide motifs that are characteristic for tubulin isotypes with pronounced cold sensitivity. The modified gene could be expressed in tandem with the appropriate counterpart of β-tubulin under control of a tuber-specific promotor such as the patatin promotor to obtain balanced and tuber-specific expression of the modified tubulin. Upon assembly into microtubules the designed tubulin would then confer increased cold sensitivity to microtubules in tuber cells and germinating buds such that higher temperatures (6-9 °C) that do not induce cold sweetening would be sufficient to suppress sprouting via disassembly of division spindles and cortical arrays that are necessary to support cell elongation in the sprouts.

Although it is unlikely that freezing tolerance of plant tissue depends on microtubules, the manipulation of the microtubular cold response has significant potential for improving the chilling resistance of growth and development. It is not necessary to create plants that can grow under the harsh conditions of Siberian winters; it might be sufficient to improve the performance of growth and leaf unfolding during the sensitive, cool period of spring in order to obtain improved usage of light energy. Even small achievements in terms of chilling sensitivity are therefore expected to produce significant increases in yield.

References

Akashi T, Shibaoka H (1987) Effects of gibberellin on the arrangement and the cold stability of cortical microtubules in epidermal cells of pea internodes. Plant Cell Physiol 28: 339-348

Akashi T, Kawasaki S, Shibaoka H (1990) Stabilization of cortical microtubules by the cell wall in cultured tobacco cells – Effects of extensin on the cold stability of cortical microtubules. Planta 182: 363-369

Anthony RG, Waldin TR, Ray JA, Bright SWJ, Hussey PJ (1998) Herbicide resistance caused by spontaneous mutation of the cytoskeletal protein tubulin. Nature 393: 260-263

Atkin RK, Barton GE, Robinson DK (1973) Effect of root-growing temperature on growth substances in xylem exudate of Zea mays. J Exp Bot 24: 475-487

Bartolo ME, Carter JV (1991a) Microtubules in mesophyll cells of nonacclimated and cold-acclimated spinach. Plant Physiol 97: 175-181

Bartolo ME, Carter JV (1991b) Effect of microtubule stabilization on the freezing tolerance of mesophyll cells of spinach. Plant Physiol 97: 182-187

Bartolo ME, Carter JV (1992) Lithium decreases cold-induced microtubule depolymerization in mesophyll cells of spinach. Plant Physiol 99: 1716-1718

Berberich T, Kusano T (1997) Cycloheximide induces a subset of low temperature-inducible genes in maize. Mol Gen Genet 254: 275-283

Berridge MJ, Irvine RF (1984) Inositol triphosphate, a novel second messenger in cellular signal transduction. Nature 312: 315-321

Bokros CL, Hugdahl JD, Blumenthal SSD, Morejohn LC (1996) Proteolytic analysis of polymerized maize tubulin: regulation of microtubule stability to low temperature and Ca^{2+} by the carboxyl terminus of β-tubulin. Plant Cell Environ 19: 539-548

Burke MJ, Gusta LV, Quamme HA, Weiser CJ, Li PH (1976) Freezing and injury in plants. Annu Rev Plant Physiol 27: 507-528

Chu B, Xin Z, Li PH, Carter JV (1992) Depolymerization of cortical microtubules is not a primary cause of chilling injury in corn (Zea mays L. cv Black Mexican Sweet) suspension culture cells. Plant Cell Environ 15: 307-312

Chu B, Kerr GP, Carter JV (1993a) Stabilizing microtubules with taxol increases microfilament stability during freezing of rye root tips. Plant Cell Environ 16: 883-889

Chu B, Snustad DP, Carter JV (1993b) Alteration of β-tubulin gene expression during low-temperature exposure in leaves of Arabidopsis thaliana. Plant Physiol 103: 371-377

Ding JP, Pickard BG (1993) Mechanosensory calcium-selective cation channels in epidermal cells. Plant J 3: 83-110

Durso NA, Cyr RJ (1994) A calmodulin-sensitive interaction between microtubules and a higher plant homolog of elongation factor 1α. Plant Cell 6: 893-905

Evans L (1975) Crop physiology. Cambridge University Press, London

Fisher DD, Cyr RJ (1993) Calcium levels affect the ability to immunolocalize calmodulin to cortical microtubules. Plant Physiol 10: 543-551

Gilmour SL, Thomashow MF (1991) Cold acclimation and cold-regulated gene expression in ABA mutants of Arabidopsis thaliana. Plant Mol Biol 17: 1233-1240

Guy CL (1990) Cold acclimation and freezing stress tolerance: role of protein metabolism. Annu Rev Plant Physiol Plant Mol Biol 41: 187-223

Haupt W (1958) Über die Primärvorgänge bei der polarisierenden Wirkung des Lichtes auf keimende Equisetumsporen. Planta 51: 74-83

Himmelspach R, Wymer CL, Lloyd CW, Nick P (1999) Gravity-induced reorientation of cortical microtubules observed in vivo. Plant J 18: 449-453

Holubowicz T, Boe AA (1969) Development of cold hardiness in apple seedlings treated with gibberellic acid and abscisic acid. J Am Soc Hort Sci 94: 661-664

Hughes MA, Dunn MA (1996) The molecular biology of plant acclimation to low temperature. J Exp Bot 47: 291-305

Irving RM (1969) Characterization and role of an endogenous inhibitor in the induction of cold hardiness in *Acer negundo*. Plant Physiol 44: 801-805

Irving RM, Lanphear FO (1968) Regulation of cold hardiness in *Acer negundo*. Plant Physiol 43: 9-13

Ishizaki-Nishizawa O, Fujii T, Azuma M, Sekiguchi K, Murata N, Ohtani T, Toguri T (1996) Low-temperature resistance of higher plants is significantly enhanced by a nonspecific cyanobacterial desaturase. Nat Biotechnol 14: 1003-1006

Ivens GW, Blackman GE (1950) Inhibition of growth of apical meristems by ethyl phenylcarbamate. Nature 166: 954-955

Jian LC, Sun LH, Lin ZP (1989) Studies on microtubule cold stability in relation to plant cold hardiness. Acta Bot Sin 31: 737-741

Juniper BE, Lawton JR (1979) The effect of caffeine, different fixation regimes and low temperature on microtubules in the cells of higher plants. Planta 145: 411-416

Kamiya N (1959) Protoplasmic streaming. Protoplasmatologia 8: 1-199

Kerr GP, Carter JV (1990) Relationship between freezing tolerance of root-tip cells and cold stability of microtubules in rye (*Secale cereale* L. cv. Puma). Plant Physiol 93: 77-82

Knight MR, Campbell AK, Smith SM, Trewavas AJ (1991) Transgenic plant aequorin reports the effects of touch and cold shock and elicitors on cytoplasmic calcium. Nature 352: 524-526

Kodama H, Hamada T, Horiguchi G, Nishimura M, Iba K (1994) Genetic enhancement of cold tolerance by expression of a gene for chloroplast Δ3 fatty acid desaturase in transgenic tobacco. Plant Physiol 105: 601-605

Krishna P, Sacco M, Cherutti JF, Hill S (1995) Cold-induced accumulation of HSP90 transcripts in *Brassica napus*. Plant Physiol 107: 915-923

Kurkela S, Franck M (1990) Cloning and characterization of a cold- and ABA-inducible *Arabidopsis* gene. Plant Mol Biol 15: 137-144

Lalk I, Dörffling K (1985) Hardening, abscisic acid, proline and freezing resistance in two winter wheat varieties. Physiol Plant 63: 287-292

Lång V, Mäntyla E, Welin B, Sundberg B, Palva ET (1994) Alterations in water status, endogenous abscisic acid content and expression of *rab18* gene during the development of freezing tolerance in *Arabidopsis thaliana*. Plant Physiol 104: 1341-1349

Lyons JM (1973) Chilling injury in plants. Annu Rev Plant Physiol 24: 445-466

Mazars C, Thion L, Thuleau P, Graziana A, Knight MR, Moreau M, Ranjeva R (1997) Organization of cytoskeleton controls the changes in cytosolic calcium of cold-shocked *Nicotiana plumbaginifolia* protoplasts. Cell Calcium 22: 413-420

Mazur P (1963) Kinetics of water loss from cells at subzero temperatures and the likelihood of intracellular freezing. J Gen Physiol 47: 347-369

Mizuno K (1992) Induction of cold stability of microtubules in cultured tobacco cells. Plant Physiol 100: 740-748

Modig C, Strömberg E, Wallin M (1994) Different stability of posttranslationally modified brain microtubules isolated from cold-temperate fish. Mol Cell Biochem 130: 137-147

Molisch H (1897) Untersuchungen über das Erfrieren der Pflanzen. Gustav Fischer Verlag, Jena, p 73

Monroy AF, Dhindsa RS (1995) Low-temperature signal transduction: induction of cold acclimation-specific genes of alfalfa by calcium at 25°C. Plant Cell 7: 321-331

Monroy AF, Sarhan F, Dhindsa RS (1993) Cold-induced changes in freezing tolerance, protein phosphorylation, and gene expression. Plant Physiol 102: 1227-1235

Monteith JL, Elston LF (1971) Microclimatology and crop production. In: Wareing PF, Cooper JP (eds) Potential crop production. Heinemann, London, pp 129-139

Morisset C, Gazeau C, Hansz J, Dereuddre (1993) Importance of actin cytoskeleton behaviour during preservation of carrot cell suspension in liquid nitrogen. Protoplasma 173: 35-47

Morisset C, Gazeau C, Hansz J, Dereuddre (1994) Is actin important for cryosurvival? Cryo-Letters 15: 215-222

Murata N, Ishizaki-Nishizawa O, Higashi H, Tasaka Y, Nishida I (1992) Genetically engineered alteration in the chilling sensitivity of plants. Nature 356: 710-713

Nick P, Schäfer E (1991) Induction of transverse polarity by blue light: an all-or-none response. Planta 185: 415-424

Petrášek J, Freudenreich A, Heuing A, Opatrný Z, Nick P (1998) Heat-shock protein 90 is associated with microtubules in tobacco cells. Protoplasma 202: 161-174

Pihakaski-Maunsbach K, Puhakainen T (1995) Effect of cold exposure on cortical microtubules of rye (*Secale cereale*) as observed by immunocytochemistry. Physiol Plant 93: 563-571

Rikin A, Richmond AE (1976) Amelioration of chilling injuries in cucumber seedlings by abscisic acid. Physiol. Plant 38: 95-97

Rikin A, Waldman M, Richmond AE, Dovrat A (1975) Hormonal regulation of morphogenesis and cold resistance. I. Modifications by abscisic acid and gibberellic acid in alfalfa (*Medicago sativa* L.) seedlings. J Exp Bot 26: 175-183

Rikin A, Atsmon D, Gitler C (1980) Chilling injury in cotton (*Gossypium hirsutum* L.): effects of antimicrotubular drugs. Plant Cell Physiol 21: 829-837

Sachs J (1865) Handbuch der Experimental-Physiologie der Pflanzen. Engelmann, Leipzig, p 514

Sakiyama M, Shibaoka H (1990) Effects of abscisic acid on the orientation and cold stability of cortical microtubules in epicotyl cells of the dwarf pea. Protoplasma 157: 165-171

Shibaoka H, Hogetsu T (1977) Effects of ethyl n-phenylcarbamate on wall microtubules and on gibberellin- and kinetin-controlled cell expansion. Bot Mag Tokyo 90: 317-321

Stair DW, Dahmer ML, Bashaw EC, Hussey MA (1998) Freezing tolerance of selected *Pennisetum* species. Int J Plant Sci 159: 599-605

Steponkus PL, Uemura M, Joseph RA, Gilmour SJ (1998) Mode of action of the *COR15a* gene on the freezing tolerance of *Arabidopsis thaliana*. Proc Natl Acad Sci USA 95: 14570-14575

Templeman WG, Sexton WA (1945) Effect of some arylcarbamic esters and related compounds upon cereals and other plant species. Nature 156: 630

Thion L, Mazars C, Thuleau P, Graziana A., Rossignol M, Moreau M, Ranjeva R (1996) Activation of plasma membrane voltage-dependent calcium-permeable channels by disruption of microtubules in carrot cells. FEBS Lett 393: 13-18

Toguri T (1996) Low-temperature resistance of higher plants is significantly enhanced by a nonspecific cyanobacterial desaturase. Nat Biotechnol 14: 1003-1006

Tucker EB, Allen NS (1986) Intracellular particle motion (cytoplasmic streaming) in staminal hairs of *Setcreasea purpurea*: effect of azide and low temperature. Cell Motil Cytoskel 6: 305-313

Warren-Wilson JD (1966) An analysis of plant growth and its control in the arctic environment. Ann Bot 30: 383-402

Wasteneys GO, Gunning BES, Hepler PK (1993) Microinjection of fluorescent brain tubulin reveals dynamic properties of cortical microtubules in living plant cells. Cell Motil Cytoskel 24: 205-213

Watson DJ (1952) The physiological basis of variation yield. Adv Agron 4: 101-145

Wolter FP, Schmidt R, Heinz E (1992) Chilling sensitivity of *Arabidopsis thaliana* with genetically engineered membrane lipids. EMBO J 11: 4685-4692.

Woods CM, Reids MS, Patterson BD (1984) Response to chilling stress in plant cells. I. Changes in cyclosis and cytoplasmic structure. Protoplasma 121: 8-16

Yuan M, Shaw PJ, Warn RM, Lloyd CW (1994) Dynamic reorientation of cortical microtubules, from transverse to longitudinal, in living plant cells. Proc Natl Acad Sci USA 91: 6050-6053

7 Tubulin Genes and Promotors

Diego Breviario
Istituto Biosintesi Vegetali, C.N.R., Via Bassini 15, 20133 Milano Italy

7.1
Summary

The main scope of plant biotechnology is to introduce desirable traits into agronomically important crops. Several transgenic crops have already become available that carry characteristics modified by transformation. Current approaches rely on the transformation with genes that are under the control of promotor sequences cloned from different heterologous sources. This approach may suffer from the low efficiency in both expression and stability of the introduced sequence. In addition, the promotor sequences that are commonly used may not be those that are best suited for the plant of interest.

This chapter will focus on an approach based on tubulin genes that may overcome some of these problems. The idea of this approach is to find a more "natural" method for plant transformation by exploiting features that are specific for coding and promotor sequences of tubulin genes. All plants contain tubulins, and tubulins can therefore be exploited to design systems for homologous plant transformation that make use of the specific characteristics of these genes. For instance, one could isolate new promotor sequences that confer tissue-specific or ubiquitous expression and use them for the production of transgenic plants that are more resistant to stress or drugs.

The characteristics of plant tubulin genes and gene products will therefore be reviewed with focus on research that could possibly lead to applications in biotechnology. In the final part of the chapter, a potential application of tubulin features for plant transformation will be discussed and a versatile molecular tubulin kit (VMTK) system for plant transformation will be proposed.

7.2
Significance of tubulin genes for agriculture

The production of hybrid lines via conventional breeding requires extensive work, and can yield plants with low viability that have often lost the advantageous traits originally present in the parent lines. In contrast, a biotechnological approach permits the introduction of new specific traits (bacterial and fungal resistance, herbicide tolerance, modified quality etc.) into cultivars that had already been selected for their agronomic value. Thus, the agronomic performance of a given cultivar is maintained and further improved due to the "surgical" introduction of additional desirable traits. To be successful, approaches that are based on transformation must use tools that are highly specific and allow for efficient regulation. Tubulin genes and promotors provide two advantages that make them prime candidates for this type of tool.

The first advantage is the presence of tubulin in all plant species, which means that tubulin represents a builtin homologous system. All plants contain a variety of tubulin promotors and coding sequences with specific patterns of expression and regulation that can therefore be exploited in biotechnological terms. A homologous approach is expected to minimize side effects. The second advantage is the reduced risk of gene silencing, which is often observed when multiple copies of a gene are transfected into the recipient plant (Flavell 1994; Matzke and Matzke 1995). Tubulin is already encoded by a gene family that has obviously found a way to overcome this problem.

From a more strictly agricultural point of view, the use of tubulin genes and promotors for plant transformation might be relevant for the following aspects:

1.   Herbicide and toxin tolerance. This approach could be an alternative to current practices based on broad-band herbicides such as glufosinate or glyphosate (Dale 1995) and should be useful to control the growth of infestants and weeds in crops that are sensitive to antimitotic drugs. A second aspect of this approach is the protection from pathogens that usurpate the microtubular cytoskeleton (such as certain plant viruses) or toxins that interfere with microtubules (see Chap. 4).
2.   Cold resistance. Cultivars of economical value but sensitive to low temperatures could gain a higher level of cold resistance in consequence of the introduction and expression of a cold-resistant tubulin isotype (see Chap. 6).
3.   Control of plant height. Shorter plants are more resistant to yield losses due to lodging and windbreak and exhibit a better partitioning between vegetative and reproductive organs (see Chap. 1).
4.   Male sterility. The introduction of male sterility into agronomically important species with a low degree of outbreeding such as rice, rapeseed, sugar beet or wheat should prove useful for producing hybrids with increased crop productivity.

To demonstrate the potential of these approaches, the characteristics of plant tubulin genes and proteins will be summarized.

7.3
Characteristics of plant tubulin genes and proteins

Tubulin is a dimeric protein constituted by a non-covalent linkage between an α- and a β-polypeptide. It is the main component of microtubules (MTs); very dynamic structures that can be organized into different arrays, in response to external signals and during different stages of plant growth and development

(Palevitz 1993; Cyr and Palevitz 1995; Wymer and Lloyd 1996). To cope with this high level of dynamics, a large set of regulatory and structural features has evolved for the tubulins (Fosket and Morejohn 1992; Goddard et al. 1994). Since microtubules are essential for cell viability and tubulin is their core element, tubulin also represents a special target molecule for a range of agents that result in serious if not lethal damage to the plant cell.

Considering the properties of tubulin from this perspective, a first target of dynamic control seems to reside in the existence of several genes for α- and β-tubulins. Plant α- and β-tubulins are encoded by two gene families, each constituted by a discrete number of different isotypes that is often higher for β-tubulins than for α-tubulins. The overall number of tubulin genes for both families is higher than that observed in animals and this is probably related to the sessile nature of plants and their need for a more complex adaptation to their environment. DNA sequencing and protein comparison data indicate that the members of the β-tubulin family originate from one single precursor, whereas two distinct subfamilies of α-tubulin (I and II) diverged early in evolution (Villemur et al. 1992, 1994).

So far, research on plant tubulin genes has mainly focused on the molecular cloning of nucleotide sequences (often cDNAs) encoding the different isotypes of α- and β-tubulin. In this way, the tubulin gene families of *Arabidopsis thaliana*, maize, rice, soybean and pea have been well characterized (Guiltinan et al. 1987; Hussey et al. 1990; Kopczack et al. 1992; Snustad et al. 1992; Liaud et al. 1992; Villemur et al. 1992, 1994; Kang et al. 1994; ; Breviario et al. 1995; KogaBan et al. 1995; Qin et al. 1997). Tubulin genes have also been isolated and characterized from other species such as potato, almond, lupin, eucalypt, *Zinnia*, carrot and others (Stocker et al. 1993; Vassilevskaia et al. 1993; Taylor et al. 1994; Carnero Diaz et al. 1996; Yoshimura et al. 1996; Okamura et al. 1997).

These reports have clearly shown that, within each tubulin family, members with a rather constitutive level of transcriptional expression coexist with (fewer) members that are characterized by specific pattern of expression and regulation. In some of these reports, changes in the relative abundance of different tubulin protein isoforms have also been documented (Ludwig et al. 1988; Bustos et al. 1989; Montoliu et al. 1989; Han et al. 1991; Joyce et al. 1992; Chu et al. 1993; Dixon et al. 1994; Gianí and Breviario 1996; Niini et al. 1996; Yoshimura et al. 1996; Okamura et al. 1997; Qin et al. 1997). This is also followed, in some reports, by analogous changes in the abundance of the respective proteins (Ludwig et al. 1988; Bustos et al. 1989; Montoliu et al. 1989; Han et al. 1991; Joyce et al. 1992; Chu et al. 1993; Dixon et al. 1994; Gianì and Breviario 1996; Niini et al. 1996; Qin et al. 1997). For instance, the *Arabidopsis thaliana TubA2* gene, the soybean *TubB2*, the potato *TubB1* and TubB2 genes and the rice *TubA2* gene are nearly constitutively expressed, whereas the *Arabidopsis thaliana TubA1*, *TubB1*, and *TubB9* genes, the soybean *TubB1* gene, the rice *TubA3* and *R1623* genes, and the maize *TubA1*, *TubA2*, *TubB1* and *TubB4* genes are preferentially expressed in specific tissues or in response to specific signals (Fig. 7.1).

PLANT TUBULINS : examples of isotype-specific expression

Isotype	_Plant_	TISSUE
TubA1	A. thaliana	mature pollen,stamens
TubB1	A. thaliana	roots
TubA1,2	maize	roots
TubB1	maize	seedling root tips
TubB4	maize	anthers,pollen
TubB1	soybean	hypocotyls
TubA3	rice	roots

Isotype	_Plant_	SIGNAL
TubB1	A. thaliana	light ▼
TubB2	A. thaliana	cold ▼
TubB9	A. thaliana	cold ▲
TubB1	soybean	light ▼
TubA3	maize	mycorrhizae,soil fungi ▲
TubA3	rice	IAA ▲

Fig.7.1. Plant tubulin isotypes known to exhibit a pattern of preferential expression in tissues or to be transcriptionally regulated by specific external signals.

These findings lead to the question as to the functional significance of this re-dundancy of tubulin genes and their differential patterns of expression. Three possible explanations have been proposed (Cleveland 1987; Goddard et al. 1994):

1. The first postulates that the redundancy of genes and the patterns of preferential expression originate from specific functional requirements. A particular microtubule array is formed in specific tissues, during specific developmental stages or under specific environmental conditions requiring the presence of specific tubulin isoforms. In *Drosophila*, this could be demonstrated nicely: the substitution of the testes-specific isoform β2 with isoform β3 resulted in the absence of several microtubule-mediated processes including meiosis and nuclear shaping (Hoyle and Raff 1990). However, similar experiments have not yet been reported for plants.

2. The second explanation emphasizes the regulatory aspects of tubulin genes and proteins. Patterns of preferential expression of specific isotypes could result from the suitability of this isotype in terms of regulation rather than function. This hypothesis predicts that the tubulin isoforms might be functionally interchangeable. The prominent presence of a specific isotype in a particular tissue or during a particular developmental stage is simply due to its preferential expression driven by the more appropriate regulatory unit.

3. The third explanation may be seen as a variation of the second. It is focused on the coevolution between specific microtubule-associated proteins (MAPs; specific for that particular tissue or condition) and their target tubulin isoform. A particular tubulin isoform is selected by its interaction with a specific set of MAPs.

So far, an extensive body of information has been accumulated on patterns of preferential isotypic expression. The functional relevance of these patterns has remained enigmatic, however. This failure would favour the regulatory explanation over the functional. The question of tubulin complexity must not be considered as a matter of academic speculation. On the contrary, its understanding is fundamental for the design of different biotechnological strategies.

There is a further aspect that is probably involved in the regulation of tubulin synthesis and accumulation, that has been neglected so far. Two different labs have reported the presence of naturally occurring tubulin antisense mRNAs in both maize and *Arabidopsis*. In maize, *TubA4* antisense mRNA has been found in those tissues where the amount of sense transcript is low (Dolfini et al. 1993). Antisense RNAs of *TubA3* was found in roots and flowers of *Arabidopsis thaliana* (Deng et al. 1996). Both findings suggest some kind of regulative role for these antisense transcripts possibly acting at the posttranscriptional level. In rice, we have also made the interesting observation that two different α-tubulin isotypes are characterized by two mRNA leader sequences that match to each other in a complementary way for a length of 67 base pairs (Qin et al. 1997).

The high level of nucleotide homology among different members of the α- and β-tubulin gene families is reflected, with few but important differences, at the amino acid level. In fact, any tubulin isotype must contain those domains that are specifically required for the assembly and disassembly of microtubules, for the interaction between the α- and β-tubulin polypeptide, and for the binding and hydrolysis of GTP (Fosket and Morejohn 1992). Some apparently minor amino acid differences can confer a different degree of sensitivity to drugs, pathogen and stress conditions to the different tubulin isotype (Fig. 7.2).

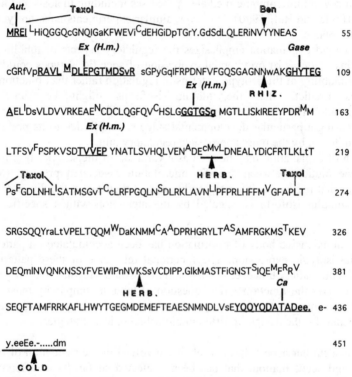

Fig.7.2. Consensus amino acid sequence obtained from the comparison among members of the β-tubulin gene family of rice, maize, soybean and *Arabidopsis thaliana*. Amino acid residues **typed in bold** and **underlined** show the domains involved in putative mRNA autoregulatory expression (**Aut.**), GTP exchange (**Ex**), GTPase activity (**Gase**), and Ca²⁺-binding (**Ca**). **Black arrows** indicate those domains which are most likely involved in the susceptible response to rhizoxin (**RHIZ**), benomyl (**Ben**), antimicrotubule herbicides (**HERB**) and to low temperatures (**COLD**). Heavy metals (**H.m.**) such as Al³⁺ are also thought to interact with tubulin by competing with Mg²⁺ for the GTP exchangeable sites (**Ex**). The presumed binding sites for taxol are also shown (Nogales et al. 1998). The consensus amino acid sequence is typed **in capitals** when the residue is fully conserved among the different plant β-tubulin members, **in lower case** when not all the compared sequences show the same amino acid residue. In this case, the amino acid more frequently present is shown. Amino acid residues **in superscript** refer to those plant residues that are different from those of their vertebrate tubulin counterparts.

Assembly and disassembly of microtubules require GTP binding and hydrolysis (Burns and Farrell 1996). Conversely, plant α- and β-tubulin polypeptides contain, in their amino acid structure, domains that are able to bind GTP (exchangeable in the case of β-tubulin, non-exchangeable in the case of α-tubulin) as well as a conserved motif putatively involved in GTPase activity (Davis et al. 1994). Mutational analyses with animal tubulins that conserve domains similar to plant tubulins have shown that GTP plays a structural role in tubulin folding (Tian et al. 1996; Zabala et al. 1996). Metal poisoning by Al^{3+} in the soil is thought to be caused by competition of the Al^{3+} with Mg^{2+} for GTP and GDP receptor sites which, in turn, would interfere severely with binding and hydrolysis of GTP (see Chap. 5).

All plant β-tubulins and subfamily II of α-tubulins contain the tetrapeptide motif MREI at their N-terminal end. This motif, which is also present in animal β-tubulins, has been shown to have a regulatory role in controlling tubulin accumulation when microtubules are poisoned with antimitotic drugs such as colchicine (Cleveland et al. 1983; Cleveland 1988). The increase in tubulin monomers caused by microtubule depolymerization or tubulin injection reduced mRNA stability, leading to a net decrease in the amount of tubulin transcript (Gay et al. 1989). However, the presence of a similar regulatory mechanism has yet not been demonstrated for plants.

Ca^{2+} binds with high affinity to animal tubulins and causes microtubule destabilization. A similar destabilizing effect of Ca^{2+} on plant microtubules has been shown recently and was discussed in relation to the modulation of cold sensitivity (Bokros et al. 1996). In this regard, several data have been obtained that correlate stability and tubulin expression in plants that were exposed to low temperature (Kerr and Carter 1990a,b; Sakiyama and Shibaoka 1990; Mizuno 1992).

Isolated plant tubulin has been shown to bind to dinitroanilines such as oryzalin with high affinity in a rapid and reversible way. These antimicrotubular drugs cause dramatic damages to plant microtubules with no apparent effect on animal microtubules (Morejohn et al. 1987; Hugdahl and Morejohn 1993; Hoffman and Vaughn 1994). Dinitroanilines are currently used to control the growth of weed grasses in crop cultivation (Holt et al. 1993). Cereals are also very sensitive to treatments with dinitroanilines. AMP is equally effective and competes with oryzalin for the binding to tubulin (Ellis et al. 1994; Murthy et al. 1994). AMP and oryzalin are much more hydrophobic than colchicine and have much higher affinities for plant tubulin. Plant cells treated with oryzalin lose their growth anisotropy as a consequence of the loss of the cortical microtubule arrays (Baskin et al. 1994). The differences in sensitivity between plant and animal microtubules to antimitotic drugs such as colchicine (reduced in plants) and herbicides (increased in plants) such as trifluralin and oryzalin are probably related to structural specificities of plant tubulin. The amino acid substitutions that have been identified as specific for the plant kingdom might play a central role in this differential drug sensitivity.

In addition to the complexity of tubulin isotypes, posttranslational modifications of animal tubulin have been shown to occur. Specific isotypes of α- and β-tubulin may become glutamylated or phosporylated. Recently, the phosporylation of the neuronal β-(III) isotype has been shown to have a role in microtubule assembly (Khan and Luduena 1996). Plant tubulin is modified posttranslationally as well and these changes further contribute to the generation of plant tubulin polymorphism (Smertenko et al. 1997). So far, there have been only a few reports providing evidence for the phosphorylation of plant tubulin. In one case, phosphorylation was associated to cold sensitivity (Koontz and Choi 1993; Mizuno 1992).

Acetylation at position 40 and tyrosination at the C-terminus of α-tubulin have also been shown. In plants, all α-tubulins are tyrosinated and the removal or the presence of the tyrosine residue as the last amino acid has been associated to a different rate of growth (Duckett and Lloyd 1994). Recently, it has been shown that gibberellic acid stimulates the acetylation of maize α-tubulin, and this correlates with a higher microtubule stability (Huang and Lloyd 1999).

Altogether the data presented in this section demonstrate the large versatility of the tubulin system. The statement seems justified that almost any aspect of gene regulation and protein binding can be addressed for tubulins. The characteristic features for the regulation of tubulin dynamics and organization are listed below:

1.   Transcriptional regulation and response to external stimuli
2.   Posttranscriptional regulation and differential mRNA stability
3.   Translational regulation and posttranslational modifications
4.   Differential protein degradation
5.   GTP binding and hydrolysis
6.   Anti-sense regulation of transcript abundancy
7.   Drug binding and interactions
8.   MAP binding
9. Ca^{2+} binding
10. Cold sensitivity

7.4
Open issues in plant tubulin research related to growth and stress responses

Two key questions are presently being addressed by current research on tubulin. The first is related to the divergent expression of different tubulin isotypes. It is not understood whether these divergent patterns reflect a functional specificity or whether they merely result from differences in promotor structure. In the second case, any sequence coding for a tubulin isotype could be replaced by other tubulin sequences under the control of a promotor that exhibits the desired pattern of tissue and developmental regulation. One has to bear in mind, however, that further levels of regulation exist as well, such as the presence of natural tubulin antisense-mRNA or the possibility of regulated mRNA stability and translation (Gonzales-Garay and Cabral 1996). The second key question concerns those differences in the amino acid sequence that distinguish animal from plant tubulin and result in pharmacologically distinct properties. An important aspect of this problem is the potential relation between tubulin structure and the resistance to cold, drugs and heavy metals.

7.4.1
Patterns of transcriptional regulation

To address the regulatory aspects, transgenic plants are produced that harbour constructs where the expression of the GUS reporter gene is driven by different tubulin promotor sequences. Northern blot analyses had shown previously that the expression of *Arabidopsis thaliana TubA1* in flowers was confined to stamens and mature pollen (Ludwig et al. 1988). This highly specific pattern of expression has been confirmed by experiments with transgenic *Arabidopsis* plants. GUS expression driven by the *TubA1* promotor was detected almost exclusively in mature and germinating pollen (Carpenter et al. 1992). These findings might indicate that the *TubA1* isotype performs a unique function in pollen germination for which other α-tubulins cannot substitute. Accordingly, an antisense approach has been launched to possibly knock out the expression of *TubA1* and to test whether the TubA1 isotype is functionally required at this stage of pollen maturation (Wick et al. 1996). Irrespective of the outcome of these experiments, the regulatory sequence of *TubA1* is a tool that can be used to confine expression of specific genes to specific stages during pollen maturation.

In tobacco plants, transgenic approaches have been undertaken to demonstrate the specific transcriptional activation of the maize *TubA3* by arbuscular mycorrhiza (Bonfante et al. 1996). The maize *TubA3* promotor sequence was fused upstream to the GUS reporter gene and β-glucuronidase activity was observed to be enhanced in cortical cells of tobacco roots that had been infected by mycorrhiza. In contrast, root cells of plants transformed with similar constructs where the *TubA3* promotor had been replaced by the maize *TubA1* promotor yielded no enzymatic activity. These data confirm that the activation of maize *TubA3* gene expression by mycorrhiza is specific for the *TubA3* promotor. Although this result shows the occurrence of a specific regulatory response, it does not exclude the possibility that, in addition, the TubA3 protein fulfills a function that is specifically required during the establishment of arbuscular mycorrhyza. This should be tested by appropriate antisense approaches. In fact, an increased expression of α-tubulin has been observed in other plants during the elevated formation of lateral roots induced by fungal colonization (Carnero Diaz et al. 1996; Niini et al. 1996).

Transgenic approaches (tubulin promotor driving expression of the GUS reporter gene) in soybean have shown that the expression of isotype *TubB1* is down-regulated by light via regulatory elements that lie within 2 kb from the translational start site of *TubB1* (Tonoike et al. 1994). The soybean *TubB1* gene had been shown previously to be highly expressed only in rapidly elongating regions of etiolated hypocotyls (Han et al. 1991). Similarily, expression of *TubB1* is specifically downregulated by light in hypocotyls of *Arabidopsis thaliana* (Leu et al. 1995).

The number of plant tubulin isotypes for which a specific pattern of expression (triggered by signals or confined to specific tissues) can be demonstrated by similar approaches is actually increasing (Chu et al. 1998; Uribe et al. 1998). In fact, similar approaches are currently performed on other plant tubulin isotypes for which a preferential pattern of expression was previously shown. Independently of the potential functional specificity of the corresponding gene products, isotype-specific tubulin promotor elements could be used to engineer constructs able to control gene expression in a time and space dependent manner.

On the other hand, several of the plant tubulin genes characterized so far show a rather constitutive pattern of transcriptional expression, i.e. the level of the respective mRNA remains more or less constant during different stages of plant growth and development or in response to external stimuli. Such a constitutive expression has been shown by a classical transgenic approach for the *TubA2* gene of *Arabidopsis thaliana* (Carpenter et al. 1993). GUS activity driven by the promotor sequence of this gene was detected virtually in all plant tissues and organs including cells with unusual microtubule arrays such as leaf trichomes, vascular tissues with developing xylem elements or pollen grains. Thus, the pattern of expression differs between the *TubA2* and the *TubA1* genes of *Arabidopsis thaliana*. This wide range of expression points to a general participation of the *TubA2* protein in all microtubule arrays that characterize the plant cell cycle. The promotor sequence of *TubA2* provides a valuable tool for biotechnological manipulation since it should confer a high and evenly distributed level of expression of any coding sequence placed under its control.

7.4.2
Regulation at the posttranscriptional level?

The development of new molecular tools must take into account three aspects of tubulin regulation. The first is related to the presence of naturally occurring antisense tubulin mRNA. The second aspect is linked to the MREI N-terminal motif that, in plants, is present in all β-tubulin sequences as well as in all members of the α-tubulin subfamily II. This motif might allow for autoregulation of tubulin mRNA through a feedback mechanisms that is triggered by increased pools of tubulin dimers in consequence of microtubule depolymerization. In principle, this feedback mechanism could be exploited to modulate the expression of a gene by altering the intracellular amount of tubulin. However, evidence for MREI-mediated tubulin autoregulation in plants is missing so far. With the use of specific α- and β-tubulin probes and antibodies, we have actually found, in rice root tips, coleoptile segments and cultivated rice cells that depolymerization of microtubules with oryzalin causes a significant decrease in the amount of tubulin protein but not in the amount of tubulin mRNA (Gianì et al. 1998). This finding suggests that the MREI regulatory mechanism may not work in plants. On the other hand, two recent reports raise the possibility that tubulin could control its own synthesis at the level of mRNA translation through the involvement of 5' and 3' UTRs (Gonzales-Garay and Cabral 1996; Anthony and Hussey 1998).

7.4.3
Structural differences between tubulin isotypes and the microtubule response to cold and drugs

Brain tubulin has been shown to be efficiently incorporated into microtubules in living plant cells (Zhang et al. 1990; Yuan et al. 1994), demonstrating that the isotypes are functionally interchangeable despite the presence of characteristic differences in the amino acid sequences. This type of experiment allows insights into general aspects of microtubular dynamics. In addition, it can be used to test whether the presence of animal tubulin can modify the properties of plant micro-tubules such as sensitivity to cold and drugs. Plant tubulin binds AMP and dini-troanilines very efficiently in contrast to animal tubulin, and the opposite is true for colchicine. Will the incorporation of animal tubulin into plant microtubules make them less sensitive to dinitroanilines? Expression studies based on the production of different isoforms of animal and plant tubulin in different systems (baculovirus, yeast) of isoforms of animal and plant tubulin should also allow detection of functional peculiarities that are caused by the presence of specific amino acid residues. In this context, the yeast system would be a good system since *Saccharomyces cerevisiae* contains only one copy of the β-tubulin gene (*Tub2*) that is essential for survival (Neff et al. 1983; Thomas et al. 1985). However, we have not been able to rescue yeast *tub2ts* mutants with the corresponding β-tubulin sequences from rice (Breviario 1997). This type of approach is principally feasible, though, as shown by the successful incorporation of a chicken-yeast chimeric β-tubulin into mouse microtubules (Bond et al. 1986). On the other hand, attempts to introduce yeast β-tubulin into the cytoskeleton of insect cells produced only low rates of incorporation (Vats-Metha and Yarbrough 1993). These findings suggest that there isotype-specific functions do exist and that different isotypes are not completely interchangeable. Nevertheless, it is still worth continuing this type of approach. The possibility of obtaining yeast cells with microtubules made up by a chimeric plant-yeast β-tubulin would allow to ask for the role of different tubulin properties that are specific for the respective plant tubulin or its yeast counterpart.

While the outcome of these experiments is to be expected in the near future, first conclusions about the relation between structural differences of tubulin isotypes and the behaviour of microtubules or even the behaviour of entire plant cells are already possible now and might be exploited.

In animal as well as in plant cells, cold tolerance has long been associated with microtubule assembly and the presence of specific tubulin isotypes. Antarctic fishes contain cold stable microtubules built up of a unique tubulin isotype different from those of fishes living in more temperate climates or bovine tubulin (Detrich et al. 1987). Quite interestingly, taxol-stabilized maize microtubules containing a β-tubulin chain where the last 15 amino acids have been deleted are more resistant to the combined destabilizing effect of Ca^{2+} and cold as compared to wild type (Bokros et al. 1996). Expression studies in *Arabidopsis thaliana* and in rye roots have shown patterns of preferential expression for certain tubulin isotypes in response to low temperature (Kerr and Carter 1990b; Chu et al. 1993). In *Arabidopsis thaliana*, those tubulin isotypes with a shorter C-terminus maintain or even increase (*TubB9*) their levels of transcriptional expression. Moreover, the presence of cold-resistant microtubules seems to confer higher tolerance to low temperatures to the whole plant (see Chap. 6). These studies indicate the possibility of working out strategies leading to production of plants that are more resistant to cold.

A second aspect is related to structural features of plant tubulin and concerns the interaction between tubulin and microtubule drugs. Plant tubulins differ consistently from animal tubulins with respect to their pharmacological properties. Data from mutants obtained in *Clamydomonas reinhardtii* (Schibler and Huang 1991), *Eleusine indica* (Vaughn et al. 1987) or *Nicotiana plumbaginifoglia* (Blume et al. 1995) with elevated resistance to microtubular herbicides indicate that amino acid substitutions in tubulin genes can confer increased resistance. Mutant lines of *Nicotiana plumbaginifoglia* with resistance to trifluralin show alterations in the pattern of β-tubulin isoforms (Blume et al. 1995). The *colR4* and *colR5* β-tubulin mutations in *Clamydomonas* exhibit increased microtubule stability and resistance to antimitotic herbicides. A corresponding missense mutation was mapped to the Lys 350 residue of β-tubulin (Schibler and Huang 1991). A hydrophobic pocket around position 200 of plant β-tubulin has long been proposed to be a site for herbicide interaction (Fosket and Morejohn 1992). Recently, two different laboratories have provided more conclusive evidence on herbicide resistance associated with mutations occurring in the α-tubulin protein (Anthony et al. 1998; Yamamoto et al. 1998). By analyzing the molecular features of a major α-tubulin isotype (*TUBA1*) present in a biotype of *Eleusine indica* (goosegrass) resistant (R) to dinitroanilines they found two amino acid substitutions with respect to the sensitive (S) biotypes. A mutation that converts threonine-239 to isoleucine was identified and shown to correlate with dinitroaniline resistance. direct proof was also provided by inserting the mutated α-tubulin gene of *Eleusine indica* into maize suspension culture cells that, in consequence, acquired resistance to dinitroanilines (Anthony et al. 1998). A second mutation involving a substitution of a methionine by a threonine at position 268 was shown to correlate to an intermediate level of resistance (Yamamoto et al. 1988). Both studies have demonstrated the feasibility of an approach by which mutated forms of tubulin, introduced either by mutagenesis or genetic crosses, can confer new advantageous features to the plant. Of course, none of these amino acid changes should fall into any of the domains involved in basic tubulin functions such as assembly or GTP binding and hydrolysis. It is therefore conceivable that alterations in these sites (introduced either by mutagenesis or by crosses with naturally occurring resistant species) could be used to introduce higher tolerance to herbicides into cultivated crops as compared to weeds.

A different type of drug, taxol, is also expected to produce interesting results. Taxol is a compound extracted from yew that has recently proved useful in the treatment of certain human cancers (Heinstein and Chang 1994). Taxol inhibits microtubule depolymerization and is often used to stabilize microtubules. One problem about taxol is that it is produced only in the bark of the yew and in a very small amount. Is this because taxol is toxic to the plant that produces it? Are yew microtubules and the corresponding tubulins more resistant to taxol than other plant counterparts? If so, is there any structural difference? These questions still await an answer and could lead to important insights on plant tubulin and related clinical applications.

Plant tubulins are sensitive to the antibiotic rhizoxin, an antimitotic substance which is produced by a fungus, *Rhizobium chinensis*, the pathogen of rice seedling blight. This pathogen threatens mainly plant roots (Koga-Ban et al. 1995). Rhizoxin binds efficiently to β-tubulin. Studies of naturally occurring mutants of *Aspergillus nidulans* that are resistant to rhizoxin as well as of resistant yeast mutants have clearly shown that sensitivity to the toxin depends on the presence of the residue Asn at position 100 of β-tubulin (Takahashi et al. 1990). This residue is maintained in the vast majority of plant β-tubulins and as yet does not appear to be involved in any tubulin function related to microtubule assembly and dynamics.

These investigations on regulatory aspects of plant tubulins and the binding of antimicrotubular drugs provide the starting point for potential biotechnological applications.

7.5
Potential approaches for manipulation

Ideally, plant biotechnological manipulations should yield plants modified for specific traits, devoid of side effects and with the acquired features possibly regulated in time and space. Manipulation of plant tubulin features could actually meet these requirements by placing an appropriate coding sequence (cold-resistant, herbicide-resistant) under the control of appropriate promotors (tissue-specific or signal-triggered). To minimize potential side effects, one could produce mutated forms of the appropriate construct. The mutations have to be introduced into those parts of the coding sequence that are located outside of the key domains known to be involved in basic tubulin functions. Among these mutations, those that do not carry side effects could be selected.

The starting point of this strategy is the central role of tubulin as the key component of microtubules. When tubulin is attacked by different agents, microtubule organization and structure are heavily affected, with drastic effects on cell viability, no matter what other effects the treatment will produce. Thus, when antimitotic substances sequester tubulin, this is enough to endanger cell division and viability. The same may be true for low temperatures, when microtubules disassemble leading to major cold-induced cellular damages.

Thus, tubulin is a good candidate for biotechnological exploitation because it provides an essential function but at the same time allows for some manipulation made possible by the intrinsic heterogeneity of its multiple isotypes.

It is therefore feasible, by recombining promotor and coding sequences, to produce "new tubulin genes" able to provide a function optimized for certain features in a specifically regulated pattern of expression (constitutive, time-dependent, tissue-specific or signal-dependent). On the basis of what is actually known, it should be possible to produce by mutagenesis new advantageous tubulin isoforms that still maintain their basic functions. This is what has actually occurred during evolution. At the beginning, one single β-tubulin gene and two α-tubulin genes provided tubulin for all cells and for all microtubular arrays. With time, plants have duplicated their α- and β-tubulin genes several times and the expression of some of them has been confined to specific tissues, to specific stages of development or in response to specific external signals. Some attention should be paid to the design of tubulin regulatory elements because the cell may not tolerate an excess of tubulin or because autoregulatory responses may occur such as translational repression mediated by UTR sequences. However, introduction of new features via the introduction of designed isotypes does not necessarily require an "overpowered" level of expression, and UTR sequences that potentially interfere with tubulin expression can be removed, as has already been demonstrated (Anthony et al. 1998). In fact, tubulin promotors have been used and shown to be active in several plant systems (Carpenter et al. 1992, 1993; Bonfante et al. 1996; Uribe et al. 1998). In principle, the creation of new combinations of promotors and modified tubulin coding sequences is no more than the continuation of the evolutionary program adopted by tubulin genes. In this way some aspects of plant growth and resistance to stress could be addressed. The VMTK (versatile molecular tubulin kits) strategy described in Fig. 7.3 could then be adopted to improve plant resistance against cold, herbicides, pathogens and heavy metals, to control plant height or to produce male sterile plants.

V.M.T.K.

Versatile Molecular Tubulin Kit

Potential exploitations of tubulin genes and regulatory sequences

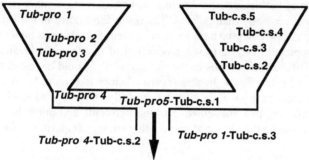

New possibly **useful combinations** and related desirable effects:

Tub-pro   Tub- coding region

Root	mutagenized tubulin
Const.	mutagenized tubulin

Herbicide resistance

Root	mutagenized tubulin

Pathogen resistance

internode	antisense-tubulin

Height control

Const.	mutagenized/C-terminus tub

Cold resistance

pollen	antisense-tubulin

Male sterility

Fig.7.3. The VMTK model. Possible new combinations between tubulin promotor (**Tub-pro**) and coding sequences (**Tub-c.s.**) are drawn together with the desirable effects that they could produce once inserted in a recipient plant. **Const.** stands for constitutively active promotor.

To illustrate the VMTK approach, specific strategies will now be briefly discussed for individual traits:

7.5.1
Herbicide resistance

This strategy has already been employed successfully using heterologous promotor sequences, giving additional credit to the VMTK approach. When the heterologous promotor can be replaced by a homologous tubulin promotor, this will represent further progress towards the production of plants where the degree and tissue pattern of herbicide resistance (e.g. in leaves, roots or other parts of the plant) could be conveniently manipulated. Thanks to the studies in the R biotypes of *Eleusine indica,* new potential sites have been identified that can be utilized for a mutagenesis approach leading to the production of plants with elevated herbicide resistance. Different mutations of the transformants should be tested to select those that have no side effects. In this regard, factors like reproductive output, fitness and inflorescence dry weight must be carefully checked. Crops with a very low rate of outcrossing that are sensitive to the herbicide and amenable to gene transformation would be the ideal material for this strategy (e.g. rice or rapeseed).

7.5.2
Rhizoxin resistance

This approach could be regarded as a subcase of the previous one. Genetic data have shown that the amino acid residue Asn100 is probably a key element for the sensitivity of microtubules to the antimitotic drug rhizoxin, released into the soil by the pathogen *Rhizobium chinensis.* Roots are particularily attacked. A feasible strategy to produce plants more resistant to this pathogen would involve the engineering of constructs with a root-specific tubulin promotor driving the expression of a tubulin isotype mutated in the amino acid position 100. Mutation of the residue 100 should not interfere with any of the basic functions carried out by plant tubulins.

7.5.3
Cold resistance

As discussed above (see also Chap. 6), plenty of data have now been collected on tubulin as a main target for the cold-induced destabilization of microtubules. Cold sensitivity seems to reside mainly in the C-terminal end of tubulin and tubulin isotypes that are more resistant to low temperatures have been characterized. Manipulation of the C-terminal end of tubulin could therefore result in the production of cold-resistant isoforms. These constructs could then be transferred into crops where yield is limited by low temperatures. The expression of this isotype could be driven by a constitutive tubulin promotor. Cold resistance might be linked to the phosphorylation of specific sites on the tubulin-coding sequence. This would allow a second approach, where these sites could then be changed to obtain cold-stable variants of tubulin.

7.5.4
Control of plant height

It is generally assumed that a reduced height is a desirable trait for many important crops (see Chap. 1). Plant height depends to a large degree on cell elongation and thus the synthesis and accumulation of tubulin. If tubulin becomes limiting, this is expected to interfere with cell elongation. This is illustrated by the response to light. Light inhibits elongation and downregulates the level of tubulin mRNA in several plants. Current field practices often disturb this natural control. The trigger for the response of elongation is the plant photoreceptor phytochrome in many cases. Thus, the overexpressing phytochrome could be used to control plant height but this approach, due to the pleiotropic effects of phytochrome, is expected to produce several unpredictable side effects (see Chap. 1). On the other hand, tubulin isotypes that are more specifically expressed in elongating internodes have been described. For the fine-tuning of tubulin expression light-regulatory elements could be inserted into the tubulin promotor sequence or antisense strategies could be designed and specifically targeted to these internode-specific tubulin isotypes. As an antisense regulation of tubulin expression seems to be occurring naturally, this approach could be used to reduce the amount of tubulin and thus to limit cell elongation. In this way, side effects should be strongly minimized.

7.5.5
Male sterility

This problem is strictly associated to the discovery of tubulin isoforms that seem to be exclusively expressed in pollen. If they perform a specific function for which no other tubulin isoform can substitute, then expression of a specific antisense mRNA driven by this pollen-specific tubulin promotor should inhibit pollen maturation and germination and should result in male sterile plants.

7.5.6
Plant transformation

The regulatory features of plant tubulin can be used to manipulate the expression of tubulin isoforms as described above; but they can also be useful for the transformation with other genes where a specific pattern of expression is desired instead of constitutive expression. For this, those tubulin promotors that are specifically active in certain tissues (i.e. root, pollen, etc.) can be used to target the expression of interesting genes to those tissues. On the other hand, tubulin promotors that are constitutively active represent a good homologous system for the constitutive expression of any heterologous gene. Constitutively active tubulin promotors can, in principle, be isolated readily by inverse PCR or conventional hybridization techniques exploiting the presence of very conservative sequences at the tubulin N-terminus. Such promotors are also expected to be relatively strong promotors, because tubulin is a highly expressed protein.

These potential applications of the VMTK system are schematically presented in Fig. 7.3. The wealth of data on different aspects of tubulin synthesis and isotype function is expected to be refined further, which should expand the versatility of the VMTK strategy that is based on two very important features of tubulin: its central role and its flexibility.

References

Anthony RG, Hussey PJ (1998) Suppression of endogenous α and β tubulin synthesis in transgenic maize calli overexpressing α and β tubulins. Plant J 16: 297-304

Anthony RG, Waldin TR, Ray JA, Bright SWJ, Hussey PJ (1998) Herbicide resistance caused by spontaneous mutation of the cytoskeletal protein tubulin. Nature 393: 260-263

Baskin TI, Wilson JE, Cork A, Williamson RE (1994) Morphology and microtubule organization in *Arabidopsis* roots exposed to oryzalin or taxol. Plant Cell Physiol 35: 935-942

Blume YAB, Kundel'chuk OP, Solodushko VG, Sulimenko VV, Yemets AI (1995) Asymmetric somatic hybrids of higher plants resistant to trifluralin. In: de Prado J, Torres G, Marshall M (eds) Proc Int Symp on weed and crop resistance to herbicides. Cordoba, Spain, April 3-6, 1995, pp 182-185

Bokros CL, Hugdahl JD, Blumenthal SSD, Morejohn LC (1996) Proteolytic analysis of polymerized maize tubulin: regulation of microtubule stability to low temperature and Ca^{2+} by the carboxyl terminus of β-tubulin. Plant Cell Environ 19: 539-548

Bond JF, Fridovich-Keil JL, Pillus L, Mulligan RC, Solomon F (1986) A chicken-yeast chimeric β-tubulin protein is incorporated into mouse microtubules *in vivo*. Cell 44: 461-468

Bonfante P, Bergero R, Uribe X, Romera C, Rigau J, Puigdomenech P (1996) Transcriptional activation of a maize α-tubulin gene in mycorrhizal maize and transgenic tobacco plants. Plant J 9: 737-743

Breviario D (1997) Rice alpha and beta tubulins: features of the gene families and regulatory aspects. Rec Res Dev Plant Physiol 1: 241-264

Breviario D, Gianì S, Meoni C (1995) Three rice (*Oryza sativa* L.) cDNA clones encoding different β-tubulin isotypes. Plant Physiol 108: 823-824

Burns RG, Farrell KW (1996) Getting to the heart of beta-tubulin. Trends Cell Biol 6: 297-303

Bustos MM, Guiltinan MJ, Cyr RJ, Ahdoot D, Fosket DE (1989) Light regulation of β-tubulin gene expression during internode development in soybean (*Glycine max* Merr.) Plant Physiol 91: 1157-1161

Carnero-Diaz E, Martin F, Tagu D (1996) Eucalypt α-tubulin: cDNA cloning and increased level of transcripts in ectomycorrhizal root system. Plant Mol Biol 31: 905-910

Carpenter JL, Ploense SE, Snustad DP, Silflow CD (1992) Preferential expression of an α-tubulin gene of *Arabidopsis* in pollen. Plant Cell 4: 557-571

Carpenter JL, Kopczak SD, Snustad DP, Silflow CD (1993) Semi-constitutive expression of an *Arabidopsis thaliana* α-tubulin gene. Plant Mol Biol 21: 937-942

Chu B, Snustad DP, Carter JV (1993) Alteration of β-tubulin expression during low-temperature exposure in leaves of *Arabidopsis thaliana*. Plant Physiol 103: 371-377

Chu B, McCune-Zierath C, Snustad DP, Carter JV (1998) Two beta-tubulin genes, *TUB1* and *TUB8* of *Arabidopsis* exhibit largely nonoverlapping patterns of expression. Plant Mol Biol 37: 785-790

Cleveland DW (1987) The multitubulin hyphotesis revisited: what have we learned? J Cell Biol 104: 381-383

Cleveland DW (1988) Autoregulated instability of tubulin mRNAs: a novel eukaryotic regulatory mechanism. TIBS 13: 339-343

Cleveland DW, Pittinger MF, Feramisco JR (1983) Elevation of tubulin levels by microinjection suppresses new tubulin synthesis. Nature 305: 738-740

Cyr RJ, Palevitz BA (1995) Organization of cortical microtubules in plant cells. Curr Opin Cell Biol 7: 65-71

Dale PJ (1995) R & D Regulation and field trialing of transgenic crops. TIBTECH 13: 398-403

Davis A, Sage CR, Dougherty CA, Farrell KW (1994) Microtubule dynamics modulated by guanosine triphosphate hydrolysis activity of β-tubulin. Science 264: 839-842

Deng WL, Haas NA, Snustad DP (1996) Characterization of naturally-occurring antisense RNAs of the *Tua3* gene in *Arabidopsis*. Plant Physiol 111: 571

Detrich HW, Prasad V, Luduena RF (1987) Cold-stable microtubules from antarctic fishes contain unique α-tubulins. J Biol Chem 262: 8360-8366

Dixon DC, Seagull RW, Triplett BA (1994) Changes in the accumulation of α- and β-tubulin isotypes during cotton fiber development. Plant Physiol 105: 1347-1353

Dolfini S, Consonni G, Mereghetti M, Tonelli C (1993) Antiparallel expression of the sense and antisense transcripts of maize α-tubulin genes. Mol Gen Genet 241: 161-169

Duckett CM, Lloyd CW (1994) Gibberellic acid-induced microtubule reorientation in dwarf peas is accompanied by rapid modification of an α-tubulin isotype. Plant J 5: 363-372

Ellis JR, Taylor R, Hussey PJ (1994) Molecular modeling indicates that two chemically distinct classes of anti-mitotic herbicide bind to the same receptor site(s). Plant Physiol 105: 15-18

Flavell RB (1994) Inactivation of gene expression in plants as a consequence of specific sequence duplication. Proc Natl Acad Sci USA 91: 3490-3496

Fosket DE, Morejohn LC (1992) Structural and functional organization of tubulin. Annu Rev Plant Physiol 43: 201-240

Gay DA, Sisodia SS, Cleveland DW (1989) Autoregulatory control of β-tubulin mRNA stability is linked to translation elongation. Proc Natl Acad Sci USA 86: 5763-5767

Gianì S, Breviario D (1996) Rice β-tubulin mRNA levels are modulated during flower development and in response to external stimuli. Plant Sci 116: 147-157

Gianì S, Qin X, Faono F, Breviario D (1998) In rice, oryzalin and abscisic acid differentially affect tubulin mRNA and protein level. Planta 205: 334-341

Goddard GH, Wick SM, Silflow CD, Snustad DP (1994) Microtubule components of the plant cell cytoskeleton. Plant Physiol 104: 1-6

Gonzales-Garay ML, Cabral F (1996) α-Tubulin limits its own synthesis: evidence for a mechanism involving translation repression. J Cell Biol 135: 1525-1534

Guiltinan MJ, Ma D, Barker RF, Bustos MM, Cyr RJ, Yadegari R, Fosket DE (1987) The isolation, characterization and sequence of two divergent β-tubulin genes from soybean (*Glycine max* L.). Plant Mol Biol 10: 171-184

Han I, Jongewaard I, Fosket DE (1991) Limited expression of a diverged β-tubulin gene during soybean (*Glycine max* Merr.) development. Plant Mol Biol 16, 225-234

Heinstein PF, Chang CJ (1994) Taxol. Annu Rev Plant Physiol Plant Mol Biol 45: 663-674

Hoffman JC, Vaughn KC (1994) Mitotic disrupters act by a single mechanism but vary in efficacy. Protoplasma 179: 16-25

Holt JS, Powles SB, Holtum JAM (1993) Mechanisms and agronomic aspects of herbicide resistance. Annu Rev Plant Physiol 44: 203-229

Hoyle HD, Raff EC (1990) Two *Drosophila* beta tubulin isoforms are not functionally equivalent. J Cell Biol 111: 1009-1026

Huang RF, Lloyd CW (1999) Gibberellic acid stabilizes microtubules in maize suspension cells to cold and stimulates the acetylation of alpha-tubulin. FEBS Lett 443: 317-320

Hugdahl JD, Morejohn LC (1993) Rapid and reversible high-affinity binding of the dinitroaniline herbicide oryzalin to tubulin from *Zea mays* L. Plant Physiol 102: 725-740

Hussey PJ, Haas N, Hunsperger J, Larkin J, Snustad DP, Silflow CD (1990) The β-tubulin gene family in *Zea mays*: two differentially expressed β-tubulin genes. Plant Mol Biol 15: 957-972

Joyce CM, Villemur R, Snustad DP, Silflow CD (1992) Tubulin gene expression in maize (*Zea mays* L.) Change in isotype expression along the developmental axis of seedling root. J Mol Biol 227: 97-107

Kang SC, Choi YJ, Kim MC, Lim CO, Hwang I, Cho MJ (1994) Isolation and characterization of two β-tubulin cDNA clones from rice. Plant Mol Biol 26: 1975-1979

Kerr GP, Carter JV (1990a) Relationship between freezing tolerance of root-tip cells and cold stability of microtubules in rye (*Secale cereale* L.cv Puma). Plant Physiol 93: 77-82

Kerr GP, Carter JV (1990b) Tubulin isotypes in rye roots are altered during cold acclimation. Plant Physiol 93: 83-88

Khan IA, Luduena RF (1996) Phosphorylation of beta (III)-tubulin. Biochemistry 35: 3704-3711

Koga-Ban Y, Niki T, Nagamura Y, Sasaki T, Minobe Y (1995) cDNA sequences of three kinds of β-tubulins from rice. DNA Res 2: 21-26

Koontz DA, Choi JH (1993) Evidence for phosphorylation of tubulin in carrot suspension cells. Physiol Plant 87: 576-583

Kopczak SD, Haas NA, Hussey PJ, Silflow CD, Snustad DP (1992) The small genome of *Arabidopsis* contains at least six expressed α-tubulin genes. Plant Cell 4: 539-547

Leu W, Cao X, Wilson TJ, Snustad DP, Chua NH (1995) Phytochrome A and phytochrome B mediate the hypocotyl-specific downregulation of *Tub1* by light in *Arabidopsis*. Plant Cell 7: 2187-2196

Liaud M, Brinkmann H, Cerff R (1992) The β-tubulin gene family of pea: primary structures, genomic organization and intron-dependent evolution of genes. Plant Mol Biol 18: 639-651

Ludwig SR, Oppenheimer DG, Silflow CD, Snustad DP (1988) The α1-tubulin gene of *Arabidopsis thaliana*: primary structure and preferential expression in flowers. Plant Mol Biol 10: 311-321

Matzke MA, Matzke AJM (1995) How and why do plants inactivate homologous (trans)genes? Plant Physiol 107: 679-685

Mizuno K (1992) Induction of cold stability of microtubules in cultured tobacco cells. Plant Physiol 100: 740-748

Montoliu L, Rigau J, Puigdomènech P (1989) A tandem of α-tubulin genes preferentially expressed in radicular tissues from *Zea mays*. Plant Mol Biol 14: 1-15

Morejohn LC, Bureau TE, Molè-Bajer J, Bajer AS, Fosket DE (1987) Oryzalin, a dinitroaniline herbicide, binds to plant tubulin and inhibits microtubule polymerization in vitro. Planta 172: 252-264

Murthy JV, Kim H-H, Hanesworth VR, Hugdahl JD, Morejohn LC (1994) Competitive inhibition of high-affinity Oryzalin binding to plant tubulin by the phosphoric amide herbicide amiprophos-methyl. Plant Physiol 105: 309-320

Neff NF, Thomas JH, Grisafi P, Botstein D (1983) Isolation of the β-tubulin gene from yeast and demonstration of its essential function in vivo. Cell 33: 211-219

Niini SS, Tarkka MT, Raudaskoski M (1996) Tubulin and actin protein patterns in Scots pine (*Pinus sylvestris*) roots and developing ectomycorrhiza with *Suillus bovinus*. Physiol Plant 96: 186-192

Nogales E, Wolf SG, Downing KH (1998) Structure of the αβ tubulin dimer by electron crystallography. Nature 391: 199-206

Okamura S, Naito K, Sonehara S, Ohkawa H, Kuramori S, Tatsuta M, Minamizono M, Kataoka T (1997) Characterization of the carrot beta-tubulin gene coding a divergent isotype, beta-2. Cell Struct Funct 22: 291-298

Palevitz BA (1993) Morphological plasticity of the mitotic apparatus in plants and its developmental consequences. Plant Cell 5: 1001-1009

Qin X, Gianì S, Breviario D (1997) Molecular cloning of three rice α-tubulin isotypes: differential expression in tissues and during flower development. Biochem Biophys Acta 1354: 19-23

Sakiyama M, Shibaoka H (1990) Effects of abscisic acid on the orientation and cold stability of cortical microtubules in epicotyl cells of the dwarf pea. Protoplasma 157: 165-171

Schibler MJ, Huang B (1991) The *colR4* and *colR15* β-tubulin mutations in *Chlamydomonas reinhardtii* confer altered sensitivities to microtubule inhibitors and herbicides by enhancing microtubule stability. J Cell Biol 113: 605-614

Smertenko A, Blume Y, Viklický V, Opatrný Z, Draber P (1997) Post-translational modifications and multiple tubulin isoforms in *Nicotiana tabacum* L. cells. Planta 201: 349-358

Snustad DP, Haas NA, Kopczak SD, Silflow CD (1992) The small genome of *Arabidopsis* contains at least nine expressed β-tubulin genes. Plant Cell 4: 549-556

Stocker M, Garcia-Mas J, Arus P, Messeguer R, Puigdomenech P (1993) A highly conserved α-tubulin sequence from *Prunus amygdalus*. Plant Mol Biol 22: 913-916

Takahashi M, Matsumoto S, Iwasaki S, Yahara I (1990) Molecular basis for determining the sensitivity of eukaryotes to the antimitotic drug rhizoxin. Mol Gen Genet 222: 169-175

Taylor MA, Wright F, Davies HV (1994) Characterization of the cDNA clones of two β-tubulin genes and their expression in the potato plant (*Solanum tuberosum* L.) Plant Mol Biol 26: 1013-1018

Thomas JH, Neff FN, Botstein D. (1985) Isolation and characterization of mutations in the β-tubulin gene of *Saccharomyces cerevisiae*. Genetics 112: 715-734

Tian G, Huang Y, Rommelaere H, Vendekerckove J, Ampe C, Cowan NJ (1996) Pathway leading to correctly folded β-tubulin. Cell 86: 287-296

Tonoike H, Han I, Jongewaard I, Doyle M, Guiltinan M, Fosket DE (1994) Hypocotyl expression and light downregulation of the soybean tubulin gene, *tubB1*. Plant J 5: 343-351

Toyomasu T, Yamane H, Murofushi N, Nick, P. (1994) Phytochrome inhibits the effectiveness of gibberellins to induce cell elongation in rice. Planta 194: 256-263

Uribe X, Torres MA, Capellades M, Puigdomenech P, Rigau J (1998) Maize alpha-tubulin genes are expressed according to specific patterns of cell differentiation. Plant Mol Biol 37: 1069-1078

Vassilevskaia TD, Ricardo CP, Rodrigues-Pousada C (1993) Molecular cloning and sequencing analysis of a β-tubulin gene from *Lupinus albus*. Plant Mol Biol 22: 715-718

Vats-Mehta S, Yarbrough L (1993) Expression of chick and yeast β-tubulin encoding genes in insect cells. Gene 128: 263-267

Vaughn KC, Marks MD, Weeks DP (1987) A Dinitroaniline-resistant mutant of *Eleusine indica* exhibits cross-resistance and supersensitivity to antimicrotubule herbicides and drugs. Plant Physiol 83: 956-964

Villemur R, Joyce CM, Haas NA, Goddard RH, Kopczak SD, Hussey PJ, Snustad DP, Silflow CD (1992) a-tubulin gene family of maize (*Zea mays* L.): evidence for two ancient α-tubulin genes in plants. J Mol Biol 227: 81-96

Villemur R, Haas NA, Joyce CM, Snustad DP, Silflow CD (1994) Characterization of four new β-tubulin genes and their expression during male flower development in maize (*Zea mays* L.) Plant Mol Biol 24: 295-315

Wick SM, Zhao KN, Li CG, Goddard RH, Eun SO, Silflow CD, Snustad DP (1996) Tubulin genes and isoforms in plants. Plant Physiol 111: 10001

Wymer C, Lloyd C (1996) Dynamic microtubules: implications for cell wall patterns. Trends Plant Sci 1: 222-227

Yamamoto E, Zeng L, Baird WV (1998) α-Tubulin missense mutations correlate with antimicrotubule drug resistance in *Eleusine indica*. Plant Cell 10: 297-308

Yan K, Dickman MB (1996) Isolation of a beta-tubulin gene from *Fusarium moniliforme* that confers cold-sensitive benomyl resistance. Appl Environ Microbiol 62, 3053-3056

Yoshimura T, Demura T, Igarashi M, Fukuda H (1996) Differential expression of three genes for different β-tubulin isotypes during the initial culture of *Zinnia* mesophyll cells that divide and differentiate into tracheary elements. Plant Cell Physiol 37: 1167-1176

Yuan M, Shaw PJ, Warn RM, Lloyd CW (1994) Dynamic reorientation of cortical microtubules from transverse to longitudinal, in living plant cells. Proc Natl Acad USA 91: 6050-6053

Zabala JC, Fontalba A, Avila J (1996) Tubulin folding is altered by mutations in a putative GTP-binding motif. J Cell Sci 109: 1471-1478

Zhang D, Waldsworth P, Hepler PK (1990) Microtubule dynamics in living dividing plant cells: confocal imaging of microinjected fluorescent brain tubulin. Proc Natl Acad Sci USA 87: 8820-8824

8 Microtubular and Cytoskeletal Mutants

Vance Baird[1], Yaroslav B. Blume[2] and Susan M. Wick[3]

[1]Horticulture Department, Clemson University, Clemson, South Carolina 29634-0375, USA

[2]Institute of Cell Biology and Genetic Engineering, National Academy of Sciences of Ukraine, Kiev, Ukraine

[3]Department of Plant Biology, University of Minnesota, St. Paul, Minnesota 55108, USA

8.1
Summary

Microtubules are biochemically one of the simplest, yet functionally most important, cellular organelles in the plant and animal kingdoms. They are integral components of a dynamic, three-dimensional framework referred to as the cytoskeleton. In addition to a fundamental role in intracellular movement, the microtubule, microfilament and intermediate filament arrays of the cytoskeleton are networks upon which asymmetric distribution of subcellular constituents is established and from which these polarized regulatory molecules mediate morphogenesis. The functions of microtubules reflect this common theme of distribution and movement. In plants, four principal microtubular functions are recognized: determination of the division plane; translocation of chromosomes; cell plate/phragmoplast formation; and control of cell morphology. Obviously, from a biotechnological point of view, the ability to control and regulate these processes via modifications of the component parts will have a profound effect on plant growth and development. Microtubules are composed primarily of a single, repeating macromolecular unit – tubulin. Tubulin itself is a heterodimeric protein, composed of two similar subunits: alpha-(α) and beta-(β)tubulin. The α- and β-tubulins are typically encoded by gene families, and these give rise to various tubulin isotypes that are differentially expressed and modified during growth and development. In addition, a ubiquitous and diverse class of proteins that bind to microtubules, known as microtubule associated proteins (MAPs), are believed to be important in nucleation, stabilization and bundling of microtubules.

Knowledge of the cytoskeleton and its fundamental role in cell morphogenesis and intracellular movement is central to an understanding of plant growth and development. Tubulin mutants have proven extremely useful for the analysis of microtubule and cytoskeletal function in algal, fungal and higher plant cells. The best-characterized mutations typically alter microtubule stability in the presence of numerous agents such as temperature, the antimicrotubular fungicides and herbicides, as well as various antitumour drugs (see Chap. 9). Tubulin gene mutations that produce non-conservative amino acid substitutions in tubulin proteins, and in turn may cause changes in the electrophoretic pattern of tubulin subunits, have been identified in *Chlamydomonas reinhardtii, Saccharomyces cerevisiae, Aspergillus nidulans, Neurospora crassa* and the higher plants *Eleusine indica* (goosegrass) and *Nicotiana plumbaginifolia*. Only goosegrass appears to possess naturally occurring site of action mutations. However, at least six other species of higher plants have been reported to have resistant biotypes, and thus may harbor similar mutations. Although these mutations can occur in either tubulin subunit, there may very well be a limited number of mutable sites resulting in phenotypes detectable as alterations in microtubule stability (e.g. sites that are critical for monomer-monomer or dimer-dimer binding). In *Chlamydomonas*, for example, single amino acid substitutions in the α- or β-tubulin subunits confer resistance to the dinitroaniline herbicides, increase microtubule stability and produce electrophoretic differences in the respective proteins as compared to wild-type tubulin. This chapter will review our current knowledge of mutant tubulins and other mutants suspected of affecting microtubule or cytoskeletal functioning in seed plants, algae and fungi. Also, it will explore and speculate on the biotechnological applications of these mutants.

8.2
Molecular components of the plant cytoskeleton

The cytoskeleton of eukaryotic cells is a study in contrasts. Firstly, the cytoskeleton is not a rigid, permanent skeleton in the strict sense, but rather a dynamic framework that is involved in establishing cell morphology and polarity rather than just cell shape maintenance (Goddard et al. 1994). Secondly, the subcellular networks that make up the cytoskeleton represent a wide range of structures and three-dimensional arrays constructed from comparatively few molecules, that display a high degree of conservation over taxonomic distance and evolutionary time. The three major structural components of the cytoskeleton are the F-actin microfilaments, the intermediate filaments and the microtubules. Plant cells have at least four basic networks in which microtubules participate either exclusively or in conjunction with other structural proteins: interphase (cortical and endoplasmic), preprophase band, spindle and phragmoplast microtubules (Lloyd 1987; Seagull 1989; Meagher and Williamson 1994; Baluška et al. 1998). The rearrangement and functioning of these networks during morphogenesis reflects the dynamic nature of microtubule stability, as well as their contribution to subcellular localization and movement. Alterations in the equilibrium between microtubule assembly and disassembly result in the success of processes as fundamental as intracellular transport and chromosome segregation.

Microtubules are biochemically one of the simplest, yet functionally most important, cellular organelles in the eukaryotic kingdoms (Lloyd 1987; Fosket 1989). The basic building block of microtubules is a polymer of tubulin. This primary tubulin unit is actually a heterodimer composed of an alpha- and a beta-tubulin polypeptide subunit. The alpha- and beta-tubulins, both composed of approximately 450 amino acid residues (M_r = 50000-55000), share on the order of

40% amino acid sequence identity and 60% nucleotide sequence identity. Also, the three-dimensional atomic structures of alpha- and beta-tubulin, as determined in vitro by electron crystallography, are nearly identical (Nogales et al. 1998). A recently constructed near-atomic model of the microtubule will help to define the molecular basis for many of the properties of these organelles (Nogales et al. 1999). The studies and information cited above point to yet another contrast: in spite of the ostensible similarity of the microtubule subunits and the apparent simplicity of microtubule structure, the dimer can assume a wide variety of conformational states (Burns 1998).

Microtubule formation is considered to be a self-assembly process (Kirschner and Mitchison 1986; Carlier 1989). The alpha- and beta-tubulin subunits assemble end-to-end to initially form the tubulin heterodimer. Then, the heterodimers polymerize, head-to-tail, in an energy dependent manner by adding predominantly to the "plus" (β-tubulin; GTP-cap) end of microtubules as they grow in a helical fashion. As visualized in the electronmicroscope, microtubules are hollow tubes, circular in cross-section (typically composed of 13 parallel oriented protofilaments), having an overall diameter of ~24 nm. However, their length can be quite variable, depending upon the particular array and stage of assembly/disassembly. Despite the numerous static images of microtubules available from every eukaryotic cell type thus far examined, it is clear that microtubule arrays continually and gradually change as cells progress through the division cycle to differentiation (Zhang et al. 1990, 1993; Erickson and O'Brien 1992; Goddard et al. 1994), and that the true image of a cytoskeletal microtubule is that of a dynamic structure in an almost constant state of flux refered to as dynamic instability (Hush et al. 1994; Yuan et al. 1994; Hepler and Hush 1996;).

Along with the well-characterized alpha- and beta-tubulins, other "tubulins" have been described. Gamma-tubulin (Oakley and Oakley 1989; Oakley 1994) shares in the order of 30% amino acid sequence identity with alpha- or beta-tubulins (Liu et al. 1994; Lopez et al. 1995). Gamma-tubulin is typically localized immunologically to the various microtubule-organizing centres present in fungal, plant and animal cells. Thus, gamma-tubulin is thought to be involved with the nucleation of microtubules. Recent genome sequencing projects focusing on baker's yeast (*Saccharomyces cerevisiae*) and the roundworm (*Caenorhabditis elegans*) have identified gamma-tubulin-like sequences in these genomes. Initial cytological characterization of the yeast protein, Tub4p, revealed its localization to the spindle pole body, as is typical for gamma-tubulin (Marschall et al. 1996). However, the nucleotide and deduced amino acid sequences are only about 30% identical to those of gamma-tubulins from other fungi, plants and animals (Burns 1995; Sobel and Snyder 1995). There is somewhat of a controversy as to whether these are bona fide gamma-tubulins or are divergent enough to be considered new members of the tubulin superfamily. In the latter case, it was proposed that that the sequence from *Caenorhabditis elegans* be termed delta-tubulin and that from *Saccharomyces cerevisiae* be termed epsilon-tubulin (Burns 1995).

Immunologically recognizable tubulin, intimately associated with cellular membranes, has been described from a number of systems including plants (Laporte et al. 1993). This so-called membrane tubulin may function in binding microtubules to membranes, transport of substances within the membrane and/or signal transduction. The uniqueness and isotypic nature of this tubulin awaits further study, as it has not yet been chemically characterized nor the encoding gene(s) cloned (Stephens 1995).

Along with the tubulins (i.e. alpha- and beta-) that make up the bulk of each microtubule, there are accessory proteins that play important roles in regulating microtubule nucleation, assembly, stability, polar organization and motility, and thus microtubule function (Cyr 1991; Wiche et al. 1991; Wymer et al. 1996). These include, among others, the microtubule-organizing centre (MTOC; centrosome) proteins (Hoffman and Vaughn 1995b; Vaughn and Harper 1998). The best-characterized of these proteins, which include centrin and various proteins recognized by monoclonal antibodies raised to mammalian cells in mitosis, is gamma-tubulin (Oakley et al. 1990; Oakley 1992; Liu et al. 1993; Marc 1997; Binarová et al. 1998). Gamma-tubulin associates with the "minus" end of microtubules, but in plants may also be associated with microtubules over their entire length. Other proteinaceous components of plant MTOCs have remained fairly elusive (Lambert 1993; Schmit et al. 1994; Stoppin et al. 1996).

Despite considerable effort, the isolation and characterization of plant microtubule-associated proteins (MAPs), which may modulate assembly and stability properties of microtubules, cross-link them via sidewall interactions or regulate the translocation of cytosolic organelles, continues to be challenging. One reason is a low degree of sequence similarity between plant and non-plant MAPs. For example, plant structural (non-motor) MAPs and some aspects of the MAP-binding domain on plant tubulin appear to be considerably different from their well-characterized counterparts in animal cells (Hugdahl et al. 1993). Nonetheless, an assortment of biochemical approaches have yielded several likely candidates (Chang-Jie and Sonobe 1993; Chan et al. 1996; Marc et al. 1996; Rutten et al. 1997). These include proteins previously determined to have roles as translation initiation and elongation factors (Durso and Cyr 1994; Bokros et al. 1995; Hugdahl et al. 1995). In addition, one energy-transducing motor MAP, shown to localize to plant mitotic spindles and phragmoplasts and to support movement toward the (minus) end of their microtubules, is a kinesin-like protein (Liu et al. 1996). Phragmoplasts also contain a microtubule translocator that is part of the bimC (blocked in mitosis) subfamily of kinesins, and it behaves much like a traditional kinesin in that it supports plus-end movement (Asada et al. 1997).

Genetic approaches to dissecting the cytoskeleton and its microtubule arrays have proven very successful, especially when coupled with biochemical and cytological analyses, for the identification of important components and for defining their physiological roles. A mainstay of genetic analysis is the identification and characterization of mutations that alter or disrupt wild-type structure and function.

These mutations can occur as primary site mutations: for example, in the tubulin genes themselves, which allows for their initial identification and further characterization. In addition, the isolation of second-site mutations (e.g. revertants or suppressors) has helped to identify components of the cytoskeleton that interact structurally with microtubules.

Molecular studies have characterized a large number of tubulin gene sequences from many different organisms, including a number of plant species (see Chap. 7, Fosket 1989; Fosket and Morejohn 1992; Liaud et al. 1992; Meagher and Williamson 1994; Breviario et al. 1995; Qin et al. 1997; Ludueña 1998). One obvious conclusion from these studies is that, at least in multicelluar plants, the tubulins are always organized in gene families, many of which can be relatively large (Kopczak et al. 1992; Snustad et al. 1992; Villemur et al. 1992; Mysore and Baird 1995; Ludueña 1998). These families give rise to various tubulin protein isotypes, and many of these are differentially expressed during growth and development (Burland et al. 1983; Silflow et al. 1987; Lewis and Cowan 1988; Fosket 1989; Hussey et al. 1991; Joyce et al. 1992; Rogers et al. 1993; Villemur et al. 1994; Chu et al. 1998; Ludueña 1998; Uribe et al. 1998; Yamamoto et al. 1998; Whittaker and Triplett 1999; Yamamoto and Baird 1999). Biochemical and immunological investigations have shown that heterogeneity in tubulin isotypes can be further enhanced by chemical modifications, which result in an increased number of tubulin isoforms incorporated into the various microtubule arrays (Wehland et al. 1984; Kozminski et al. 1993; Hoffman and Vaughn 1995a; Smertenko et al. 1997a). These posttranslational modifications can occur in the form of phosphorylation, acetylation, reversible tyrosination/detyrosination, polyglutamination (and formation of D2-tubulin) or polyglycylation (MacRae 1997; Smertenko et al. 1997b). Besides differential gene expression and the question of biological relevance of the various modifications in plant systems, an important area of investigation centres on the functional significance of the numerous isotypes/isoforms (e.g. are they biologically neutral, are they adaptive, do they mediate unique microtubular functions?).

Another general conclusion from comparative nucleotide and amino acid sequence analyses is that the alpha- and beta-tubulins represent ancient gene duplications that diverged from a common ancestral gene, which predated the divergence of the four eukaryotic kingdoms (Fosket and Morejohn 1992). That is to say, on average, the alpha-tubulins from one species are more similar to the alpha-tubulins of any other species than they are to the beta-tubulins from that same original species. Furthermore, there is evidence that individual isotypes within the alpha-tubulin family may represent ancient lineages conserved over evolutionary time (Villemur et al. 1992), as is known for plant actins (Hightower and Meagher 1986). Interestingly, while the C-terminus of individual beta-tubulins appears to be conserved among distantly related animals (Cleveland 1987), the last 20 amino acids of plant beta-tubulins provide little indication of conserved lineages. Despite the range in tubulin isotypes outlined above, it has been shown that tubulins from plants, animals, fungi and protists will copolymerize (see for example Hepler and

Hush 1996). Taken together, such homogeneity and conservation is most likely the result of evolutionary constraints on sequence divergence placed upon the tubulins by other interacting molecules (e.g., during either protein-ligand and/or protein-protein binding)

Knowledge of the microtubule cytoskeleton and its fundamental role in cell morphogenesis and intracellular movement is central to an understanding of plant growth and development. Microtubule mutants provide insight and useful tools for the analysis of cytoskeletal function in several evolutionarily diverse groups; e.g. algae, fungi and seed plants (Oakley 1985). Such mutants were selected for altered phenotypes following chemical- or irradiation-induced mutagenesis, while others occur naturally in populations and have been selected in the laboratory or inadvertently through agricultural practices. The best-characterized mutations are those whose phenotype shows altered microtubule stability in the presence of numerous agents such as temperature, antimicrotubular fungicides and herbicides, and various antitumour drugs (see Chap. 9). The mode of action of most of these mitotic disrupters is thought to involve direct binding to free tubulin heterodimers (Hess and Bayer 1977; Bajer and Mole-Bajer 1986; Morejohn et al. 1987; Hugdahl and Morejohn 1993; Ellis et al. 1994; Hoffman and Vaughn 1994; Murthy et al. 1994). These drug-tubulin complexes are then added to the plus-end of microtubules, but they inhibit further polymerization of dimers or drug-dimer complexes. Therefore, continued microtubule growth is halted. The process of dynamic instability then results in the rapid disappearance of microtubules from the various cytoskeletal arrays, influenced by inherent differences in stability of each array (e.g. cortical microtubules are typically more stable than are spindle microtubules).

Naturally occurring or mutagenesis-induced point mutations that may cause alterations in the electrophoretic pattern of tubulin isotypes have been identified in *Chlamydomonas reinhardtii*, *Saccharomyces cerevisiae*, *Aspergillus nidulans*, *Neurospora crassa* and the seed plants *Eleusine indica* and *Nicotiana plumbaginifolia* (Thomas et al. 1985; Orbach et al. 1986; Lee and Huang 1990; Jung et al. 1992; James et al. 1993; Baird et al. 1996; Anthony et al. 1998; Blume et al. 1998; Yamamoto et al. 1998). In addition, other species of plants (such as *Amaranthus palmeri*, *Setaria viridis*, *Sorghum halepense*, *Poa annua*) have been reported to have herbicide-resistant biotypes and thus may harbour similarly interesting and important mutations.

This chapter will review current knowledge of mutant tubulins and other mutants suspected of affecting microtubule or cytoskeletal functioning in plants, but will include important examples from algae and fungi. It will also explore and speculate on the biotechnological applications of these mutants.

8.3
Mutants in yeasts, other fungi and slime moulds

The fungi and slime moulds have played an important role in our fundamental understanding of the cytoskeleton (Huffaker et al. 1987; Oakley 1999). For example, the study of model yeast systems, phytopathogenic fungi and *Physarum*, and their responses to antifungal agents have provided an important source of interesting tubulin mutants, and even led to the identification of a new class of tubulins. Mutants resistant to the antimicrotubule benzimidazoles (Davidse and Flach 1977; Kilmartin 1981) appeared in the 1970s following periods of intense and/or exclusive use of these cost-effective, systemic fungicides (Delp 1980; Davidse 1986). It is assumed that mutants resistant to the N-phenylcarbamates, which also display an antimicrotubule mode of action, arose through nearly identical processes.

Research efforts over the past decade have catalogued and characterized a large number of mutant alleles that confer different degrees of resistance or sensitivity to the antimicrotubule fungicides. Many of these mutants mapped to loci known to encode tubulin genes; drug resistance tends to be linked to beta-tubulin mutations, while increased sensitivity is traced to mutations in alpha-tubulin genes (Sheir-Neiss et al. 1978; Morris et al. 1979; Neff et al. 1983; Umesono et al. 1983; Burland et al. 1984; Hiraoka et al. 1984; Toda et al. 1984; Orbach et al. 1986; Foster et al. 1987; May et al. 1987; Schatz et al. 1988). These mutants provide insight into microtubule functioning in that their mode of action disrupts fundamental cytoskeletal processes. For instance, *Aspergillus* beta-tubulin mutants that are supersensitive to benomyl are unable to support normal nuclear transport during conidial germination (Oakley and Morris 1980). In contrast, a benomyl-resistant mutant shows blocked chromosome movement due to microtubule disruption, or hyperstabilization of microtubules resulting from hampered microtubule disassembly (Oakley and Morris 1981; Oakley 1985).

Many of these mutants have been further characterized at the molecular level. These efforts identified specific lesions (i.e. point mutations) and correlated them with the resistance phenotypes characterized from field isolates or selected in laboratory strains (Thomas et al. 1985; Orbach et al. 1986; Jung and Oakley 1990; ; Machin et al. 1995; Cruz and Edlind 1997). Studies of a dozen or more filamentous and non-filamentous fungal species have identified at least nine mutable codons. The more frequently altered and biologically interesting codons are those at residues 6, 198 and 200 (along with 165, 167, 238 and 241) of a consensus fungal beta-tubulin protein molecule. Most of these are within β-strands in the amino-terminal domain (GTP-binding) of the polypeptide (Nogales et al. 1998). Conversion of His-6 to Tyr or Leu converts susceptible strains to resistant ones (Jung et al. 1992). The same is true for mutations of codon 198 (that convert Glu to Ala, Lys, Gly, Gln or Asp) and codon 200 (converting Phe to Tyr) (Jung et al. 1992; Koenraadt et al. 1992; McKay and Cooke 1997). However, the level of resistance can vary, depending on the residue encoded. For example, Ala or Lys at codon 198 confers high levels of resistance to the benzimidazoles, while Gly or Glu is reported to yield moderate or intermediate levels of resistance. Similarly, Tyr at codon 200 confers a level of drug resistance comparable to that of the latter cases (Koenraadt et al. 1992; Koenraadt and Jones 1993; McKay and Cooke 1997). This pattern is reflected in (or predicted by) the pharmacological differences between fungal, animal and plant beta-tubulins. Mammals are naturally tolerant of the benzimidazoles, and their beta-tubulins typically have a Tyr at a codon comparable to Phe-200 of fungi (and thus analogous to the moderately resistant fungal mutants). Plants, which are highly resistant to these drugs, typically have either methionine or serine at this position. It can be suggested that such a mutation in a fungal beta-tubulin would create strains with greater resistance than have yet been observed.

Where studied, and not unexpectedly, these mutations result in pleiotropic effects consistent with the strain's fungicide/drug resistance phenotype. This is to say, besides showing reduced affinity for the chemical agent (and related agents), as documented by *in vitro* binding studies, mutant strains display enhanced microtubule stability in the presence of the drug (both in vitro and in vivo) and show alterations in *in vivo* processes requiring microtubular components such as mitosis, cytokinesis, nuclear migration or morphogenesis (Raper 1966; Morris 1986; Kamada et al. 1989).

Interestingly, a number of benzimidizole-resistant strains of, for example, *Aspergillus nidulans*, *Botrytis cinerea*, *Venturia inaequalis* and *Neurospora crassa*, display (super)sensitivity to the N-phenylcarbamates and vice versa (Jones et al. 1987; Fujimura et al. 1992b; Koenraadt et al. 1992; Koenraadt and Jones 1993). This phenomenon is termed negatively correlated cross-resistance, and a preliminary model to explain the data suggested that altering the polar nature of certain amino acid residues along the beta-tubulin molecule alters the protein affinity for various subclasses of fungicides (Fujimura et al. 1990). For example, mutations that affect codon 198 converting Glu to Gly or Ala will reduce the binding affinity of beta-tubulin for carbendazim while simultaneously increasing it for diethofencarb. In fact, a sensitivity series has been demonstrated for the *Saccharomyces cerevisiae* beta-tubulin with Gly yielding maximal sensitivity that decreased over Ala, Lys, and the wild-type Glu (producing the maximal resistance). Interestingly, and in contrast to the work with filamentous fungi reported above, the mutant yeast strains were identical in their level of resistance to benomyl (Fujimura et al. 1992a).

One goal of evaluating mutants at the molecular level is to identify and assign functions to the various domains in the tubulin protein. Continued efforts in *Saccharomyces cerevisiae* have proven successful at providing such insight. For example, Reijo et al. (1994) performed a systematic mutational analysis of the beta-tubulin gene (i.e. targeted charged amino acids converted to Ala). Three quarters of the 55 constructed mutations were not lethal to cell growth, despite the singularity of the gene and its obvious sequence conservation. The lethal substitutions were located primarily in three regions, presumably defining domains most critical for beta-tubulin function. Such random (if not prolific) production of tubulin mutants suggests the possibility of creating specific mutations, as alluded to earlier, to investigate microtubule functioning by characterizing their phenotypic effects. Along these lines Sage et al. (1995) introduced a point mutation in the beta-tubulin gene (TUB2) of the budding yeast, *Saccharomyces cerevisiae*. They created a Thre to Lys substitution at codon 107. The tub2-T107K mutant produced microtubules with slowed growth and catastrophic-disassembly rates in vitro, which in turn slowed cell-cycle progression in vivo. These phenotypic effects correlated with the mutation site located in a region of the beta-tubulin protein proposed to be involved with GTP binding and its assembly-dependent hydrolysis. Li et al. (1996) used site-directed mutagenesis and gene replacement to artificially create a yeast strain with a mutation in the beta-tubulin gene at codon 167, resulting in a Phe to Tyr substitution. This mutant strain displayed a three- to four- fold decrease in sensitivity to carbendazim and nocodazole, as seen previously from work with *Neurospora crassa* (Orbach et al. 1986). However, the mutant strain also showed an eightfold increase in sensitivity to benomyl. These findings suggest that benzimidazoles interact with a specific site on beta-tubulin, namely, that amino acid residue 167 interacts with position-1 of the benomyl molecule (Li et al. 1996).

The various fungal systems described above are also amenable to the identification of other genes, which can enhance our understanding of the originally described mutations and thus cytoskeletal functioning in general. An important subclass of such mutations are those resulting in suppression of the original mutant phenotype and thus reversion to a wild-type phenotype. Especially interesting are those that are trans to the original mutations (i.e. extragenic suppressor mutations vs. back mutations) (Oakley and Morris 1981; Morris 1986; Kamada et al. 1990). Isolation and subsequent analyses of *Aspergillus nidulans* strains harbouring suppressor mutations of a beta-tubulin mutation, *benA33*, led to the genetic identification of a number of mutants that mapped to the alpha-tubulin locus (Morris et al. 1979; Oakley and Morris 1981; Oakley et al. 1987). This same approach also led to the important discovery of entirely new members of the tubulin superfamily, the gamma-tubulins (Oakley and Oakley 1989). Similar investigations in *Saccharomyces cerevisiae* showed that two of the over 30 conditional lethal mutants that mapped throughout the locus of the major, essential alpha-tubulin gene (*TUB1*) could be reverted to wild type due to extragenic suppressor mutations that occurred in the single beta-tubulin gene, *TUB2* (Schatz et al. 1988). In contrast, an earlier screen for non-complementing mutants, unlinked to this beta-tubulin gene, identified a mutation that mapped to the major alpha-tubulin locus (Stearns and Botstein 1988). Despite the hope that such an approach would identify non-tubulin genes that encode physically interacting components of the microtubule system, only mutations that mapped to the three known tubulin genes were isolated. On the other hand, a similar approach in *Chlamydomonas reinhardtii* was successful in characterizing new mutant alleles of *APM1* and *APM2*, which affect resistance to amiprophosmethyl and oryzalin and thus flagellar assembly/regeneration. These mutations did not map to any of the four tubulin gene loci (James and Lefebvre 1992).

Tubulin genes from the fission yeast, *Schizosaccharomyces pombe*, have been isolated through similar studies using conditional lethal mutants that affect microtubule-mediated processes (e.g. nuclear division, migration and fungicide sensitivity) only under specialized conditions, and that can be complemented (Toda et al. 1983; Umesono et al. 1983). Two mutations affecting fission yeast cell cycle were used to identify tubulin genes. By complementation of *nda2* (*ben1*) mutants, which alter microtubular organization and/or produce supersensitivity to the benzimidazoles, wild-type clones for the two alpha-tubulin genes were isolated (Toda et al. 1984). A second mutation, *nda3*, which is cold-sensitive, lacks spindles at the restrictive temperature and is supersensitive to thiabendazole, led to the cloning of an allele of the single beta-tubulin gene (Hiraoka et al. 1984). Although the molecular nature of the conditional lethal mutants is unknown, the *nda3* complementing gene, *NDA3*, which is derived from a thiabendazole resistant strain (*ben1-TB101*), contains Leu at codon 6, instead of the His typically found at that position in wild-type beta-tubulins of other fungi. As mentioned above, such an amino acid substitution is correlated with resistance to the anti-microtubule benzimidazoles.

In *Saccharomyces cerevisiae*, another interesting mutant tubulin was recovered (Archer and Vega 1995). The polypeptide encoded by this alpha-tubulin allele (tub1-724) binds with lower affinity to beta-tubulin than does the wild-type polypeptide (Vega et al. 1998). This, in turn, results in heterodimer dissociation and an excess of free beta-tubulin monomer, which is lethal (Katz et al. 1990). The molecular basis of this mutant alpha-tubulin polypeptide is the loss of a positive charge at amino acid residue 106. It is speculated that Arg in the wild-type allele participates in phosphate binding of the non-exchangeable GTP, and thus indirectly functions in α–β binding during heterodimer formation (Vega et al. 1998). Yeast cells can be rescued from beta-tubulin lethality (Weinstein and Solomon 1990). One way is by the production of an excess of wild-type alpha-tubulin. Cells can also be rescued by the production of an excess of the beta-tubulin-binding protein Rbl2p. Therefore, the investigation of this mutant has allowed the further characterization of proteins (e.g. cofactors) involved in tubulin heterodimer formation and microtubule assembly: e.g. Rbl2p, Rki1p, Pac10p and a tubulin-binding protein, Pac2p, that aids folding into an assembly competent state (Alvarez et al. 1998; Archer et al. 1998; Smith et al. 1998). Although these and other cofactors may not be required for cell viability, at least one CIN1 homologue from fission yeast, alp1, appears to be essential and was shown to associate with microtubules in vivo (Hirata et al. 1998).

8.4
Mutants in algae

Photosynthetic protists are also important experimental systems for the genetic and biochemical characterization of microtubule arrays in flagella (axonemes), mitotic spindles and the cortical cytoskeleton (Seagull 1989; Lechtreck and Melkonian 1991: Hoffman and Vaughn 1995b), and some of the first mutants directly affecting cytoskeletal structure and function were isolated from the unicellular green alga *Chlamydomonas reinhardtii*. These mutants were selected from mutagenesis screenings because they were resistant to antimicrotubular drugs. They were associated with altered beta-tubulin isoforms, paralleling the findings in fungi (; Sheir-Neiss et al. 1978; Thomas et al. 1985; Orbach et al. 1986) and animal tissue culture cell lines (Cabral et al. 1980). Subsequently, a further mutation that confers resistance to microtubule-inhibiting drugs was localized to an alpha-tubulin gene.

Two mutants, *colR4* and *colR15*, were initially identified by their ability to grow clonally on media containing 5 mM colchicine. They were shown by 2D-PAGE and in vitro translation to express beta-tubulin proteins altered in their primary amino acid sequence, rather than altered by posttranslational modification (Bolduc et al. 1988). The mutations mapped to a single genetic locus on chromosome XII, in one of the two beta-tubulin genes (which encode identical proteins in *Chlamydomonas reinhardtii*).

Sequence analysis of the tubulin genes from the two mutants revealed the occurrence of two different point mutations in the same conserved codon (Lys-350) of the beta-2-tubulin gene (Lys-350). In *colR4*, an adenine is exchanged by a guanine base leading to a substitution of the wild-type basic Lys at codon 350 by an acidic Glu residue. In the *colR15* variant, on the other hand, the Lys is replaced by an apolar Met as the result of an adenine to thymidine base transversion (Lee and Huang 1990). These mutations, resulting in β-tubulin isoforms with increased acidic charge, account nicely for the altered migration that had been originally observed in IEF and 2D-PAGE (Bolduc et al. 1988). Lysine-350 is conserved among many plant and animal β-tubulins, but differs from that found in yeast and other fungi, where it is replaced by either Glu or Leu.

Further analysis of the mutant strains suggested that resistance is a consequence of altered tubulin-tubulin interactions that enhance assembly and/or stability of the microtubules. In addition to colchicine, these two mutants were shown to be cross-resistant to a number of different microtubule-disrupting agents such as vinblastine, oryzalin or amiprophosmethyl, as well as exhibiting an increased sensitivity to the microtubule-stabilizing alkaloid taxol (paclitaxel). Although residue 350 lies adjacent to one of the two Cys implicated in colchicine binding (Little and Ludueña 1985), the missense mutations in *colR4* and *colR15* probably affect a more fundamental aspect of microtubule stability rather than just the drug-binding specificity of the mutated beta-tubulin molecule (Bolduc et al. 1988; Lee and Huang 1990). In addition to these mutants, Schibler and Huang (1991) investigated a number of spontaneous revertants, argued to be secondary intragenic mutations, and showed that in each strain the pattern of drug sensitivity cosegregated and coreverted with the beta-tubulin mutations. This study provided the first genetic evidence that the in vivo phytotoxic effects of antimicrotubulare herbicides are actually related to microtubule function. Using immunoblotting and immunofluorescence, the mutants were shown to display differences in the cellular pattern of acetylated alpha-tubulin as compared to that of the wild type. The authors concluded that expression of the mutant beta-tubulins confers altered sensitivity to various inhibitors of microtubule activity by enhancing the stability of those microtubules that contain the mutated beta-tubulin. In support of this hypothesis, the atomic model of the tubulin dimer localizes Lys-350 near the non-exchangeable GTP binding site of the alpha-tubulin (Nogales et al. 1998) – a position important for the correct alignment of the alpha- and beta-tubulin subunits within the heterodimer.

Similar conclusions regarding microtubule stability were reached for the tubulin mutation *upA12* that had been produced in the strain *apm1-18* of *Chlamydomonas reinhardtii* by means of ultraviolet light mutagenesis. The strain *apm1-18* is characterized by a five- to sixfold increased resistance to amiprophosmethyl and oryzalin (James et al. 1988), and the *upA12* mutation alone caused an additional twofold increase in resistance to amiprophosmethyl and oryzalin. Interestingly, this resistance to these drugs that block assembly of microtubules was accompanied by a twofold greater sensitivity to paclitaxel (James et al. 1993), a drug that inhibits disassembly of microtubules. The mutation was not linked to *apm1-18*, but rather mapped to a centromere-proximal interval on linkage group III containing the alpha-1-tubulin gene (*TUA1*), one of the two alpha-tubulin gene loci that encode identical proteins. Immunoblotting and *in-vitro* translation showed that a non-acetylated electrophoretic variant of alpha-tubulin was present in *upA12*, and that this mutant isoform cosegregated with drug resistance. DNA sequence analysis revealed that the molecular lesion was a single missense mutation. It occurred in a highly conserved, hydrophobic region of the alpha-tubulin molecule between the amino acid residues 20 and 30, just outside the H1 helix predicted from molecular modelling. The base transition (T to C) results in a change in codon 24 from Tyr (TAC) in the wild-type allele to His (CAC) in the *upA12* allele (*tua1-1*). This non-conservative amino acid replacement, of a basic residue for an uncharged polar residue was consistent with the electrophoretic mobility shift observed on two-dimensional immunoblots of axonemal proteins.

Interestingly, the *upA12* mutant strain displays four α-tubulin isoforms on 2D-PAGE, consisting of two primary gene products (the wild-type and the mutant isoforms) and two modified isoforms (the acetylated wild-type and the acetylated mutant isoforms). The non-acetylated mutant isoform was not detected in microtubules of the axoneme, and thus may be defective in assembly, whereas the other three isoforms were localized in this organelle. This indicates that the tua1-1 monomer represents an improved substrate for acetylation (and possibly other post-translational modifications), and that the increased stability of axonemal microtubules in the mutant may be at least in part related to this improved competence for acetylation. The paclitaxel supersensitivity of *upA12* is consistent with the interpretation that the mutation affects fundamental aspects of microtubule stability. Furthermore, molecular modelling shows that Tyr-24 is located adjacent to Thr-239 (Nogales et al. 1998), a position thought to be important for correct dimer-dimer interactions (see the following section on seed-plant mutants). This does not rule out an effect on ligand binding, and an alteration of a site important for herbicide binding may also play a role in the drug-response phenotype, since the *upA12* mutant displays unaltered sensitivity to colchicine.

8.5
Mutants in seed plants

The microtubular cytoskeleton of seed plants, particularly angiosperms, has been the subject of numerous investigations (for reviews see Fosket 1989; Lloyd 1991; Fosket and Morejohn 1992; Cyr 1994; Goddard et al. 1994; Meagher and Williamson 1994; Shibaoka and Nagai 1994; Assaad et al. 1997). The majority of these studies have described the cytoskeleton, characterized its components and described various functions using wild-type systems. Although natural variation between species (e.g. monocot versus dicot species) has contributed to our knowledge, the genetic approach (for instance the analysis of naturally occurring or artificially induced mutants) can enhance and refine our understanding.

8.5.1
Mutants with alterations of cell division

Although many proteins important for cell division, and in particular those associated with microtubule function, are now known from plant systems such as structural or fibrous MAPs or energy-transducing motor proteins (Durso and Cyr 1994; Hugdahl et al. 1995; Chan et al. 1996; Gu and Verma 1996; Liu et al. 1996; Asada and Collings 1997; Gu and Verma 1997) have been identified in plants, such systems have not as yet proven particularly amenable to mutational and molecular analyses (Liu and Meinke 1998; Traas and Laufs 1998). However, embryo development, with its apparently strict dependence on highly ordered cell division and expansion, has long been a process targeted by investigators in search of answers to fundamental questions of morphogenesis and development (Meinke 1986; Meinke 1991; Sheridan and Clark 1993; Sheridan 1995; Raghavan 1997; Liu and Meinke 1998). Therefore, the identification of genes that regulate embryogenesis and subsequent organogenesis is likely to provide new tools to study the control of cytoskeletal structure and function.

8.5.1.1
GNOM

In this regard, mutants initially described in terms of their effect on differentiation and gross morphology may represent lesions in proteins essential to cytoskeletal functioning. One example is the *GNOM* gene from *Arabidopsis thaliana*. This gene and others were identified, because certain non-lethal mutations that were produced by chemical mutagenesis significantly altered embryo and/or seedling body organization without causing a premature arrest of embryogenesis (Jürgens et al. 1991). The phenotypic effect of different mutations in *GNOM* is that of an axial (i.e. apical-basal) pattern deletion (Mayer et al. 1991), and *GNOM* was shown to be important at the earliest stages of embryogenesis. For example *gnom* zygotes exhibit an almost symmetrical first division, resulting in two nearly equal-sized cells, whereas this division is typically asymmetric in wild-type zygotes (Mayer et al. 1993). This is followed by further chaotic divisions of the apical cell, such that gnom seedlings lack a root and a hypocotyl. Furthermore, cotyledons may be reduced or eliminated. The deduced GNOM protein shares homology with the products of two yeast genes; *YEC2*, a non-essential gene, and *SEC7*, which encodes a cytosolic secretory protein localized in the Golgi apparatus (Shevell et al. 1994; Busch et al. 1996). Thus, *GNOM* is unlikely to represent a primary structural component of the cytoskeleton. However, it does appear to be important in partitioning information, and *GNOM* is hypothesized to be active as a multimer, possibly functioning in protein secretion (Shevell et al. 1994; Busch et al. 1996).

8.5.1.2
MONOPTEROS

Similar to *GNOM*, the *MONOPTEROS* gene regulates the establishment of proper apical-basal patterning in developing embryos of *Arabidopsis thaliana* (Jürgens et al. 1991; Mayer et al. 1991). Mutant *monopteros* embryos display abnormalities in cell arrangement as early as the octant stage. They have four cell tiers, rather than the two tiers of the wild type, and the alignment of cell walls is often irregular, indicating that the MONOPTEROS gene product is required early in embryogenesis. By the heart stage, mutant embryos display altered development of lower-tier cell derivatives. Inner cells of mutant embryos are less elongate, more isodiametric, show a more random pattern of cell division and do not produce the cell files characteristic of wild-type embryos. Also, as a result of the disturbed pattern of cell division, mutant embryos fail to form correctly the uppermost suspensor cell, the hypophysis, leading to impaired development of the seedling root. This can culminate, for strong alleles of *monopteros*, in the elimination of hypocotyl and radicle, including the root apical meristem. However, these alleles may or may not alter the number and arrangement of the cotyledons, and the shoot apical meristem is always normal in both structure and function. It was speculated that the mutant phenotypes (at least those of the strong mutants) most likely result from complete gene inactivation. Therefore, because cellular defects associated with monopteros mutants are restricted to the basal cells of the embryo and the uppermost suspensor cell, it is proposed that wild-type *MONOPTEROS* gene activity is absolutely required for basal element formation (Berleth and Jürgens 1993). The *MONOPTEROS* gene (*IAA24*) was recently isolated by positional cloning, and six mutants were characterized (Ulmasov et al. 1997; Hardtke and Berleth 1998). Five of the mutants are nonsense mutations creating stop codons within exons 10 or 11, and the sixth is a frame shift (deletion) mutation in exon 11. The MONOPTEROS protein is predicted to be a transcription factor since it contains both a DANN-binding domain and nuclear localization signals. Because of sequence similarities, the binding domain is predicted to bind *cis*-acting control elements in promoters of auxin-inducible genes.

8.5.1.3
KNOLLE and KEULE

KNOLLE, an other gene that is important for correct embryogenesis, affects cell division directly. As with *GNOM* and *MONOPTEROS*, the *KNOLLE* gene was originally identified from investigations into the mechanisms of embryonic pattern formation in *Arabidopsis thaliana* (Mayer et al. 1991), but unlike these two mutant types, the effects of *knolle* alleles were interpreted in terms of radial patterning. Mutations in *KNOLLE* interfere with the normal rate and plane of cell division. Mutant alleles, each leading to gene inactivation, produce defects in the setup of radial layers during early embryogenesis, and their effect is evident by the eight-cell stage. These defects result in a globular embryo in which a well-formed, wild-type epidermis is morphologically lacking. The outer cell layer of *Knolle* embryos does not grow as a distinct sheet because the cells do not uniformly divide in the anticlinal plane. Rather, the orientation of the divison plane varies, resulting in both anticlinal and periclinal divisions, as well as in incomplete cross-walls protruding from the peripheral parental wall and abnormal cell enlargement. The predicted KNOLLE protein shows similarities to syntaxins (Bassham et al. 1995), membrane-anchored proteins that function as receptors for vesicle-docking molecules. Furthermore, KNOLLE has been immunologically localized to the cell plate (Lauber et al. 1997). Thus, it is postulated that KNOLLE is involved in vesicle trafficking during cytokinesis, most likely mediating membrane fusion events in the phragmoplast (Lukowitz et al. 1996; Lauber et al. 1997).

Certain aspects of the *Knolle* phenotype can be phenocopied in different species by a treatment with drugs such as caffeine (Gunning 1982; Hepler and Bonsignore 1990). For example, incomplete cell plates are formed in root meristem cells of faba bean following treatment with caffeine, and the dividing cells produce only outgrowths from the parental side-walls instead (Röper and Röper 1977). This situation mimics that in the *Arabidopsis* mutant *Keule*, where all populations of dividing cells and their progeny in the early embryo show misoriented division planes and have incomplete cross-walls (Assaad et al. 1996). Recent studies of cytokinesis in the alga *Spirogyra* revealed the existence of two separate but coincidental mechanisms to accomplish cell division. If microtubules are disrupted by drug treatments, the actin-based cleavage process still initiates cytokinesis from the cell periphery. This results in an incomplete cross-wall with a gap at its centre (McIntosh et al. 1995).

A phenotype reminiscent of *Knolle* and *Keule* is observed in the pea mutant *cytokinesis-defective* (*cyd*). This mutation, which is inherited as a recessive lethal, alters normal cell plate formation to produce "stubs" attached to the side walls. Interestingly, the phenotype is observed only in non-meristematic cells starting at the late globular/early heart stage, and can be mimicked in wild type pea seedlings by treatments with caffeine (Liu et al. 1995). The *CYD* gene product is suggested to function in vesicle transport and/or be related to proteins of the actin-myosin system.

Although the biochemical and molecular explanation for these results is currently lacking, the data are consistent with the existence of a fundamental regulatory step in cell division involving vesicle transport and fusion, and points to the importance of a genetic approach in order to elucidate the true mechanism.

8.5.1.4
FASS

A phenotype with altered cell divisions could be produced by recessive mutations in a further gene regulating morphogenesis (Torres-Ruiz and Jürgens 1994). Like *KNOLLE*, the *FASS* gene product appears to affect the pattern of cell division. However, in contrast to *KNOLLE*, mutants of *FASS* are not considered pleiotropic. Thus, *fass* seedlings are able to differentiate a full complement of cell, tissue and organ types, despite the highly irregular pattern of cell division characteristic of fass embryos. Like *GNOM*, *FASS* is required in the very early stages of embryogenesis through to adult plants, but unlike *GNOM*, the *FASS* gene does not affect cell polarity – as evidenced by *Fass* zygotes having a normal, asymmetrical first division yielding unequal daughter cells that stain differentially. *FASS* plays a key role during subsequent cell expansion and additionally affects the orientation of new cell walls. It has been suggested that *FASS* is required for the correct organization of cortical microtubules in transverse or helical arrays (Traas and Laufs 1998), because *Fass* (as does the related *tonneau-1*) mutants of *Arabidopsis* lack microtubular preprophase bands (Traas et al. 1995; McClinton and Sung 1997). Alternatively, the primary role of *FASS* has been related to the regulation of auxin metabolism, because strong *FASS* alleles could not be rescued by the application of gibberellin (Torres-Ruiz and Jürgens 1994), and phenocopies of *Fass* mutants could be produced by cultivating wild-type embryos in a medium containing the stable auxin NAA (Fisher et al. 1996).

On the other hand, cell division defects observed for phenotypically similar, *Fass*-like mutations in maize (*tangled-1*) and *Arabidopsis* (*tonneau-1*) may prove to have a cytoskeletal basis for their mechanism of action (Traas et al. 1995; Smith et al. 1996; Mazars et al. 1998). TANGLED-1, for instance, is believed to be required for positioning of the cytoskeletal arrays that establish the division plane in longitudinally dividing cells, and for spatial guidance of expanding phragmoplasts (Cleary and Smith 1998). Because the cells of *tonneau-1* mutants, similarly to *Fass*, have randomly arranged microtubules at interphase and lack preprophase bands, TONNEAU-1 is suggested to control the organization of the transverse or helical arrays of cortical microtubules (Traas and Laufs 1998). Full explanations await further cytological and molecular analyses.

8.5.1.5
Other genes involved in cell division

The functions of the genes discussed above are clearly associated with morphogenesis at the cellular level. Despite their apparently ubiquitous presence in plants, they seem to play an indirect role in cell division. The polarity, orientation and timing defects displayed by the various mutants suggest that the translation products of these genes that had originally been interpreted as elements of pattern formation may interact with a more fundamental component(s) of the cytoskeleton, which regulates its assembly and structure.

In addition to the genes discussed here, a large collection obviously exists of genes involved in the successful completion of the division cycle. Many of them were identified from mutants with impaired progression through mitosis or meiosis and cytokinesis in yeast and mammalian systems. Genes that are important for DNA replication, phosphorylation cascades, or the maintenance of the meristematic condition have been isolated from plants as well. These include histone genes, the cell division cycle genes, such as cyclin (CLN) and cyclin-dependent kinase (CDK) homologues, homeobox genes and cytokinesis-defective mutants. Although some affect cytoskeletal functions as well, their effect seems to be indirect and they will not be discussed here (for review see Francis et al. 1998).

8.5.2
Microtubule and tubulin mutants in the strict sense

An early investigation into the components of the cytoskeleton of plants studied microtubule stability in seedlings of carrot (*Daucus carota*), due to its unique properties of being naturally tolerant to trifluralin, yet having small, lipid-depauperate seeds. Using electron microscopy and immunofluorescence, it was shown that microtubules of the four major arrays in root cells are not disrupted by treatments with saturated solutions of most dinitroaniline herbicides (Vaughan and Vaughn 1988). Carrot was also shown to be cross-resistant to two other mitotic disrupters, amiprophosmethyl and hexanitrodiphenylamine. The biochemical and molecular explanation for the resistance of carrot to herbicides at concentrations that disrupt microtubules in both tolerant (e.g. cotton and soybean) and sensitive species (e.g. maize and sorghum) has not been elucidated, but it does not seem to involve a generalized detoxification or compartmentalization of the herbicides.

Despite repeated attempts using chemical mutagenesis to select mutants resistant to the phytotoxic effects of antimicrotubule drugs in *Arabidopsis thaliana* (Smeda and Vaughn 1994), the first microtubule mutants in higher plants were those from agricultural environments. Since the early 1970s, the preemergence dinitroaniline herbicides have been used extensively to control small-seeded and monocotyledonous weeds in row crops, as well as in vegetables and turfgrass. The continued and repeated use of these inexpensive and effective herbicides (e.g. Pendulum/pendimethalin, Treflan/trifluralin, or Surflan/oryzalin) resulted in the selection of weed biotypes with genetic resistance to these antimicrotubular mitotic disrupters. The first report of naturally occurring resistance was in *Eleusine indica* (goosegrass) from the Southeastern United States (Mudge et al. 1984; Murphy et al. 1986).

These plants proved to be cross-resistant to all dinitroanilines and the chemically unrelated organophosphorus phosphorothioamidate herbicides (such as amiprophosmethyl). However, the resistant biotype was found to be sensitive to the disruptive effects of colchicine, vinblastin and pronamide, and even supersensitive to carbamate herbicides and griseofulvin (Mudge et al. 1984; Vaughn et al. 1987). There is some controversy regarding the sensitivity of the resistant mutant biotype to the microtubule-stabilizing effects of paclitaxel, as compared to that of the susceptible biotype (Vaughan and Vaughn 1987; Anthony and Hussey 1999). Cytological studies of roots using mitotic indices and electron microscopy showed that the resistant R biotype is able to maintain functional microtuble arrays (i.e. cortical, spindle and phragmoplast) in the presence of concentrations of antimicrotubule drugs that would disrupt these arrays in the susceptible/sensitive S biotype (Vaughn and Koskinen 1987; Vaughn et al. 1987; Vaughn and Vaughan 1986). It was speculated that the high level of resistance observed in the R-biotype (i.e. 0.1 mM oryzalin, trifluralin and amiprophosmethyl) indicated a site-of-action mutation, rather than increased production of the target molecule, herbicide decomposition or a decrease in effective herbicide concentration.

A second resistant biotype was also described from agricultural fields in South Carolina (Mudge et al. 1984). This biotype is considered intermediately resistant (I) to trifluralin, and shows pharmacological and cytological differences from the R-biotype (Vaughn et al. 1990). Because the I-biotype displayed differential resistance to the various dinitroanilines tested, and because phragmoplast microtubules were more sensitive than spindle or cortical microtubules, it was hypothesized that the I-biotype has a unique mechanism of resistance different from that of the R-biotype (Vaughn et al. 1990).

Following these reports in goosegrass, a number of other weed species subjected to trifluralin selection have been shown to be resistant to the dinitroanilines and various other herbicides and drugs with similar modes of action. These include green foxtail (*Setaria viridis*), Palmer amaranth (*Amaranthus palmeri*), Johnsongrass (*Sorghum halepense*), rigid ryegrass (*Lolium rigidum*), black-grass (*Alopecurus myosuroides*), barnyard grass (*Echinochola crus-galli*) and annual bluegrass (*Poa annua*) (Morrison et al. 1989; Moss 1990; Gossett et al. 1992; James et al. 1995; McAlister et al. 1995; Nikolova and Baeva 1996; Heap 1997; Yelverton and Isgrigg 1998). The specificity and/or level of resistance to the anti-microtubule dinitroanilines varied among these species. However, the broadest cross-resistance and highest level of resistance is displayed by the R-biotype of goosegrass. Insight into the mechanism of antimicrotubule resistance in the mutant R-biotype of goosegrass was obtained from the analysis of isolated tubulin protein. Studies of purified tubulins extracted from both the S- and R-biotypes showed that, although they polymerized into microtubules in the absence of oryzalin, only tubulin isolated from the R-biotype was able to form recognizable microtubules in the presence of this dinitroaniline herbicide (Vaughn and Vaughan 1990). This information combined with other data indicated that tubulin of the R-biotype may be fundamentally different from that of the S-biotype by either altered herbicide binding and/or the production of hyperstable microtubules.

Genetic studies were initiated in three dinitroaniline-resistant weed species. In *Setaria viridis*, resistance was determined to be inherited as a single, recessive nuclear gene (Jasieniuk et al. 1994). Although it was shown that this recessive resistance could be transferred from *Setaria viridis* to cultivars of foxtail millet (*Setaria italica*), the data suggested a more complicated mechanism with two linked loci and part of the resistance being under the control of a number of minor genes (Wang et al. 1996). There was speculation that resistance in goosegrass would be controlled by a gene that is completely or partially dominant (Jasieniuk et al. 1994), or even be quantitative in its inheritance (Smeda and Vaughn 1994). However, genetic evidence proves that dinitroaniline resistance is inherited in a simple Mendelian fashion as a single, recessive nuclear gene, but with three alleles at that locus (Zeng and Baird 1997, 1999).

Despite preliminary reports of biotype-specific restriction fragment length polymorphisms and electrophoretic variants of tubulin, early molecular and biochemical studies reported in the literature failed to conclusively identify DNA polymorphisms or qualitative or quantitative differences in tubulin proteins that correlated with the dinitroaniline herbicide response phenotypes (Waldin et al. 1992; Mysore and Baird 1995); neither was there evidence for gene amplification as a possible explanation for herbicide resistance. Furthermore, large DNA deletions/insertions or rearrangements were ruled out, and it was postulated that if resistance involved a site-of-action mutation, it would then most likely be a point mutation (Mysore and Baird 1995).

Subsequent analysis focused on cloning the various alpha- and beta-tubulin genes in search of mutations that might explain the differential sensitivities to dinitroaniline exposure. These efforts failed to identify any missense mutations in four beta-tubulin genes (Yamamoto and Baird 1999), suggesting that the resistance in the higher plant goosegrass differed from that in *Chlamydomonas* and the benzimidazol/N-phenylcarbamate-resistant yeasts and other fungi. In agreement with this were reports of an alpha-tubulin from the R-biotype that was altered in its electrophoretic properties as compared to the predominant species in root tips of the S-biotype (Waldin 1995; Baird et al. 1996). Reported changes in charge and molecular weight implied that the mutation would more likely reside in the alpha-tubulin gene, rather than the isoform resulting from posttranslational modification. Recent studies succeeded in identifying mutations in two of the three alpha-tubulin genes (Anthony et al. 1998; Yamamoto et al. 1998). However, only the mutations in the *alpha-1 tubulin* gene (*TUA1*) strictly correlated with the genetic data from F_1-hybrids and F_2 plants from crosses between the three biotypes (i.e. S x R, S x I and I x R) (Zeng and Baird 1997, 1999; Yamamoto et al. 1998). In these cases, homozygous susceptible plants had only the wild-type *TUA1* (S-)allele, and homozygous highly or intermediately resistant plants had only the R-allele or I-allele, respectively. Most importantly, both the F_1 and F_2 heterozygote plants had the predicted combination of S, I and R alleles at the *Tua1* locus.

The missense mutations in *TUA1* affected two sites, both coding for conserved amino acids, in the intermediate domain of the alpha-tubulin (Nogales et al. 1998). One is a highly conserved Thr in the H7 helix, and the other a conserved Met in the B7 β-strand. In the R-biotype, codon 239 of *TUA1* has a nucleotide transition that converts the wild type Thr codon (ACA) into a relatively rare codon coding for Ile (ATA) (Anthony et al. 1998; Yamamoto et al. 1998). This modification increases the hydrophobicity of the alpha-tubulin polypeptide, because it replaces a polar (hydrophilic) residue with a non-polar (hydrophobic) residue, and predicts a change in secondary structure around the mutated site from a helix to a β-strand. The atomic structure model predicts this position to be near the region of dimer-dimer interaction within the protofilament. Therefore, these alterations may simultaneously change the electrophoretic mobility and the drug-binding or dimer-stability properties of the molecule (Yamamoto et al. 1998).

In the I-biotype, on the other hand, a mutation in codon 268 predicts a substitution of the wild-type Met (ATG) by a Thr (ACG). This would cause the hydrophobicity of the polypeptide to decrease only slightly and not alter its secondary structure (Yamamoto et al. 1998). These differences may be reflected in the different levels and mechanisms of resistance conferred by the two alleles. Using transgenic calli, it was proven that the mutant alpha-tubulin R-allele (*drpr*) can confer resistance to antimicrotubular drugs. Maize Black Mexican Sweetcorn calli expressing a cDNA of this mutant gene were shown to be resistant to oryzalin and trifluralin at 0.1 mgl⁻¹, a concentration that clearly distinguishes them from wild-type callus and callus expressing the susceptible allele (*DrpS*) (Anthony et al. 1998). The degree of resistance depends on copy number. Using immunofluorescence microscopy and epitope tagging, the authors confirmed that the resistant alpha-tubulin transgene product is incorporated into cortical, spindle and phragmoplast microtubule arrays. These findings were then repeated in a regenerable plant system using transgenic tobacco and 0.4 mgl⁻¹ pendimethalin, oryzalin or amiprophosmethyl (Anthony et al. 1999).

In goosegrass, the three alleles can be aligned in a dominance series with *DrpS* > *drpi* > *drpr*. For example, heterozygous plants, *DrpS/drpi* or *drpi/drpr*, display the phenotype of the "most-suscpetible" parent (i.e. S or I, respectively) rather than showing an additive level of resistance. This observation, and the existence of multiple alleles raised an interesting question as to whether an alpha-tubulin allele that was mutated in two sites would confer a greater level of resistance than the naturally occurring highly resistant homozygous recessive *drpr/drpr*. This was answered recently using transformed maize calli, where it could be shown that the double mutant increased the level of tolerance to dinitroaniline and phosphoro-thioamide herbicides by a value nearly equal to the summation of the resistance levels conferred to calli expressing either single mutation alone (Anthony and Hussey 1999). This further supports the hypothesis that each mutation exerts its effect through a unique process. This work clearly implies that through natural recombination, as well as through recombinant DNA manipulations, even greater levels of resistance can be obtained.

Recently, *in vitro* mutagenesis and selection of protoplasts of *Nicotiana plumbaginifolia* on media containing a dinitroaniline or a phosphorothioamide herbicide resulted in the isolation of mutant lines (Yemets et al. 1997b; Blume et al. 1998). These experiments were the first to demonstrate that tubulin mutants may be developed artificially (Blume and Strashnyuk 1993). A number of mutant lines display cross-resistance to trifluralin and amiprophosmethyl. The best-studied lines are seven- to tenfold more resistant to these antimicrotubule drugs than is the wild-type parental line, as evidenced by analysis of microtubule stability in isolated protoplasts using immunofluorescence microscopy. Genetic analyses concluded that trifluralin resistance is inherited as a semidominant nuclear encoded trait, while amiprophosmethyl resistance is dominant (Blume et al. 1996 1998).

Two-dimensional polyacrylamide gel electrophoresis (2-D PAGE), followed by immunoblotting of tubulins isolated from trifluralin- or aminprophosmethyl-resistant lines, established that in every case the resistant lines expressed an electrophoretic variant of beta-tubulin as compared to the control wild-type line. The molecular weight of the mutant beta-tubulin isoform did not appear to differ from that of the wild-type beta-tubulins, but it possessed a more acidic pI. Therefore, as with the goosegrass alpha-tubulin, the molecular lesion(s) in the *Nicotiana plumbaginifolia* beta-tubulin may be more likely to be a mutation in the primary sequence of the beta-tubulin gene, rather than result from a post-translational modification.

The analysis of the mutant beta-tubulin was extended in studies involving the transfer of antimicrotubule drug resistance using somatic hybridization. Asymmetric hybrids were created using leaf mesophyll protoplasts from the trifluralin-resistant line *tfl12r* or the amiprophosmethyl-resistant line (*apm5r*) fused to protoplasts of susceptible *Nicotiana sylvestris* or *Atropa belladonna* (Yemets et al. 1997a, 2000). Resistant hybrids were selected and regenerated in vitro on media containing the selective drug concentration. Cytogenetic and inheritance analyses confirmed the presence of genetic material (from one to three chromosomes) transferred from the resistant lines in the asymmetric hybrids, and 2-D PAGE analysis revealed the expression of an additional beta-tubulin in the asymmetric somatic hybrids, corresponding to the mutant isoform acquired from the resistant *Nicotiana plumbaginifolia* parent. These results support the hypothesis that the existence of microtubule arrays that are stable in the presence of antimicrotubule drugs, in the parental lines and in the somatic hybrids, is the result of genetic resistance conferred by a mutant beta-tubulin. Preliminary sequence data from partial cDNAs for four beta-tubulins expressed in leaves of an *Atropa belladonna* x *Nicotiana plumbaginifolia* (*apm5r*) hybrid indicate that Lys-350 is conserved in these transcripts, and thus resistance may differ from that characterized in the case of the *colR4* and *colR15* strains of *Chlamydomonas reinhardtii*.

As an alternative to studying naturally occurring or induced missense mutations insertional mutations of tubulin genes can be produced and analyzed using transposable elements and T-DNA insertions. This approach to examining functions of specific tubulin genes should be especially useful in plants in which all tubulin genes have been sequenced already. Identification of plants containing tubulin gene insertional mutations involves PCR screening of insertion libraries, using a tubulin gene primer and an insert primer. While insertions into the 5' or 3' UTRs may alter timing and location of tubulin gene expression, inserts into coding regions are likely to produce gene knockouts. Mutations that are gametophytically lethal will be lost after the first generation, and knockouts that result in the loss of a major tubulin isotype may also be lethal if the plant cannot compensate by producing more of the other isotypes to maintain normal levels of total alpha and beta tubulins. However, insertional mutants with more subtle or conditional phenotypes may shed light on the question of why so many genes exist for alpha, beta, and gamma tubulins in plants. Knockouts of beta-tubulin genes could be particularly informative because, unlike the alpha-tubulins, beta-tubulin isovariants tend to be uniquely different from each other, both within and among species. That is, they tend not to be grouped into subfamilies of related proteins, nor do they appear to be part of ancient gene families, with a particular isotype having an obvious counterpart in other species, based on sequence similarities.

8.6
Biotechnological implications

Antimicrotubular drugs can function as strong selection agents. For the development of basic transgenic systems, especially for graminaceous species, the mutant tubulin genes conferring antimicrotubule drug resistance may be appropriate to be used as selectable markers. The mutant tubulin genes would function much like the typical antibiotic resistance marker currently employed for selection. The presumptive transformed cells/tissue would be placed in a culture medium containing the antimicrotubule drug. The non-transformed cells that are susceptible to the incorporated drug, would be unable to maintain normal cell division and thus would be eliminated. Only the transformed cells, expressing the mutant tubulin gene(s), would be viable under the selective conditions. A number of refinements (such as appropriate promoters driving single or combinations of alpha- and beta-tubulins) and restrictions to certain plant species (e.g. those without natural resistance to dinitroanilines, phosphorothioamidates or carbamates) will have to be met before such selectable markers could be put into routine use. However, variations in the resistance phenotype, for either the level or type of drug tolerated, that is expressed as a result of the specific mutation can provide additional levels of selection.

An obvious application for the characterized mutant tubulin genes is sought in the development of herbicide-resistant crops through biotechnological methods (Yemets and Blume 1999). The early steps in this process have been realized recently. Patrick Hussey's group successfully expressed, in maize callus culture, a mutant alpha-tubulin originally isolated from a graminaceous weed displaying high levels of resistance to the dinitroanilines (Zeng and Baird 1997, 1999; Anthony et al. 1998; Yamamoto et al. 1998). This success was quickly followed by transgenic experiments in regenerable lines of tobacco (Anthony et al. 1999). The widespread extension of such technology to other crops will depend upon a number of factors; not the least of which will be the economic and industrial future of this class of herbicides, and the need to prevent the transfer of resistance traits to non-target species through natural hybridization (i.e. weedy, wild relatives of the transgenic crop).

Conversely, new pesticides could be designed and developed that are more efficient, environmental compatible and cost-effective. Such efforts could be based upon our knowledge gained from understanding the function and mechanism of the resistance conferred by the mutant tubulins (Cai and Cresti 1999). Comparisons of nucleic acid and biochemical data from existing mutants with the emerging molecular model of the tubulin dimer and assembled microtubules are improving our understanding of the resistance mechanism. However, the available mutations cover only a small portion of the mutable sites in alpha- and beta-tubulin. Studies like those of Reijo et al. (1994), Sage et al. (1995), Li et al. (1996) and S.M. Wick, (unpublished results) using mutagenesis in an attempt to saturate the tubulins with site-directed point mutations or transposon insertions will provide new material for analysis and reveal domains likely to be important in microtubule function and cytoskeletal dynamics.

Along with a more complete understanding of the mechanisms underlying resistance and susceptibility gained from studying mutant and wild-type tubulins, existing antimicrotubule drugs can be extended into new systems. An example of such an application is the use of herbicides to inhibit infections by human parasites (Chan and Fong 1990, 1994; Chan et al. 1993). Recent studies have shown that trifluralin can function as an effective antiprotozoan drug, at least *in vitro*, where proliferation of the protozoan parasite is preferentially inhibited. This result is most likely due to the similarity of tubulins from trypanosomatids (e.g. *Plas-*

modium falciparum, Leishmania mexicana and *Trypanosoma brucei*) to those of plants and their difference from mammalian tubulins. However, commercial applications await further analysis and the development of dinitroanilines with reduced toxicity in mammalian systems (Bell 1998).

With more and more investigations of pollen formation, pollen-tube growth, and pollen-stigma interactions, the importance of the cytoskeleton in sexual reproduction has become quite evident. Althoug further information from the genetic and molecular level is needed, it is possible to speculate on areas of research that may lead to biotechnological applications (Cai and Cresti 1999). Two of these areas are improvements of in-vitro fertilization systems and the induction of male sterility through manipulation or perturbation of the cytoskeletal apparatus. As discussed earlier, it is clear that an active cytoskeleton is fundamental for the development and growth of the embryo from the zygote. It seems likely that the cytoskeleton will also be important in mediating gamete fusion and zygote formation. Control of these activities may eventually allow for the routine production of sexual hybrids between diverse taxa. Similarly, targeted disruption of the pollen tube cytoskeleton could provide generic or species-specific methods for inducing male sterility using either a genetic or an epigenetic approach.

Finally, as more is learned about plant MAPs, these proteins may have future biotechnological applications (even beyond the plant kingdom). MAPs have been postulated to function as cytosolic anchors and thus they may participate in the development and maintenance of subcellular localization and polarity of morphogens and stored messages. The ability to control or engineer this capability has implications for plant development both in vivo and in vitro. A mammalian MAP, tau, functions to stabilize microtubular tracks; however, in patients with Alzheimer's disease, the degree of phosphorylated tau protein is elevated above normal levels. In neuronal cells, an increase in tau concentration was shown to impair the transport of organelles known to be transported along microtubule tracks (Ebneth et al. 1998). Potentially, information gained from studies in plant systems could impact the field of human mental disorders.

References

Alvarez P, Smith A, Fleming J, Solomon F (1998) Modulation of tubulin polypeptide ratios by the yeast protein Pac10p. Genetics 149: 857-864

Anthony RG, Hussey PJ (1999) Double mutation in *Eleusine indica* α-tubulin increases the resistance of transgenic maize calli to dinitroaniline and phosphorothioamidate herbicides. Plant 18: 669-674

Anthony RG, Waldin TR, Ray JA, Bright SWJ, Hussey PJ (1998) Herbicide resistance caused by spontaneous mutation of the cytoskeletal protein tubulin. Nature 393: 260-263

Anthony RG, Reichelt S, Hussey PJ (1999) Dinitroaniline herbicide-resistant transgenic tobacco plants generated by co-overexpression of a mutant α-tubulin and a β-tubulin. Nat Biotechnol 17: 712-716

Archer J, Magendantz M, Vega L, Solomon F (1998) Formation and function of the Rb12p-β-tubulin complex. Mol Cell Biol 18: 1757-1762

Archer JE, Vega LR (1995) Rbl2p, a yeast protein that binds to b-tubulin and participates in microtubule function *in vivo*. Cell 82: 425-434

Asada T, Collings D (1997) Molecular motors in higher plants. Trends Plant Sci 2: 29-37

Asada T, Kuriyama R, Shibaoka H (1997) TKRP125, a kinesin-related protein involved in the centro-some-independent organization of the cytokinetic apparatus in tobacco BY-2 cells. J Cell Sci 110: 179-189

Assaad FF, Mayer U, Wanner G, Jurgens G (1996) The *KEULE* gene is involved in cytokinesis in Arabidopsis. Mol Gen Genet 253: 267-277

Assaad FF, Mayer U, Lukowitz W, Juergens G (1997) Cytokinesis in somatic plant cells. Plant Physiol Biochem 35: 177-184

Baird WV, Morejohn L, Zeng L, Mysore K, Kim HH (1996) Genetic, molecular and biochemical characterization of dinitroaniline herbicide resistance in goosegrass (*Eleusine indica*). In: Brown H, Cussans GW, Devine MD, Duke SO, Fernandez-Quintanilla C, Helwig A, Labrada RE, Landes M, Kudsk P, Streibig JC (eds) 2nd Int Weed Control Congress, Copenhagen, Denmark, April 3-9, 1996, pp 551-557

Bajer AS, Mole-Bajer J (1986) Drugs with colchicine-like effects that specifically disassemble plant but not animal microtubules. Ann N Y Acad Sci 466: 767-784

Baluška F, Barlow PW, Lichtscheidl IK, Volkmann D (1998) The plant cell body: a cytoskeletal tool for cellular development and morphogenesis. Protoplasma 202: 1-10

Bassham DC, Gal S, Conceicao ADS, Raikhel NV (1995) An *Arabidopsis* syntaxin homologue iso-lated by functional complementation of a yeast *pep12* mutant. Proc Natl Acad Sci USA 92: 7262-7266

Bell A (1998) Microtubule inhibitors as potential antimalarial agents. Parasitol Today 14: 234-240

Berleth T, Jürgens G (1993) The role of the monopteros gene in organising the basal body region of the *Arabidopsis* embryo. Development 118: 575-587

Binarová P, Hause B, Dolezel J, Dráber P (1998) Association of γ-tubulin with kineto-chore/centromeric region of plant chromosomes. Plant J 14: 751-757

Blume YB, Strashnyuk NM (1993) Obtaining microtubule protein mutants. Cytol Genet (Tsitologiya i Genetika) 27: 78-92

Blume YB, Strashnyuk NM, Solodushko VG, Smertenko AP, Gleba YY (1996) Alterations of ß-tubulin confers resistance to trifluralin of *Nicotiana plumbaginifolia* mutants obtained in vitro. Proc Natl Acad Sci Ukraine (Russ) No 7: 132-137

Blume YB, Strashnyuk NM, Smertenko AP, Solodushko VG, Sidorov VA, Gleba YY (1998) Altera-tion of ß-tubulin in *N. plumbaginifolia* confers resistance to amiprophos-methyl. Theor Appl Genet 97: 464-472

Bokros CL, Hugdahl JD, Kim HH, Hanesworth VR, Van Heerden A, Browning KS, Morejohn LC (1995) Function of the p86 subunit of eukaryotic initiation factor (iso)4F as a microtubule-associated protein in plant cells. Proc Natl Acad Sci USA 92: 7120-7124

Bolduc C, Lee VD, Huang B (1988) ß-tubulin mutants of the unicellular green alga *Chlamydomonas reinhardtii*. Proc Natl Acad Sci USA 85: 131-135

Breviario D., Giani S, Meoni C (1995) Three rice cDNA clones encoding different β–tubulin isotypes. Plant Physiol 108: 823-824

Burland TG, Gull K, Schedl T, Boston RS, Dove WF (1983) Cell type-dependent expression of tubu-lins in *Physarum*. J Cell Biol 97: 1852-1859

Burland TG, Schedl T, Gull K, Dove WF (1984) Genetic analysis of resistance to benzimidazoles in *Physarum*: differential expression of β–tubulin genes. Genetics 108: 123-141

Burns RG (1995) Identification of two new members of the tubulin family. Cell Motil Cytoskel 31: 255-258

Burns RG (1998) Synchronized division proteins. Nature 391: 121-123

Busch M, Mayer U, Jürgens G (1996) Molecular analysis of the *Arabidopsis* pattern formation gene *GNOM*: gene structure and intragenic complementation. Mol Gen Genet 250: 681-691

Cabral F, Sobel ME, Gottesman MM (1980) CHO mutants resistant to colchicine, colcemid, or grise-ofulvin have an altered β–tubulin. Cell 20: 29-36

Cai G, Cresti M (1999) Rethinking cytoskeleton in plant reproduction: toward a biotechnological future? Sex Plant Reprod 12: 67-70

Carlier MF (1989) Role of nucleotide hydrolysis in the dynamics of actin filaments and microtubules. Int Rev Cytol 115: 139-170

Chan J, Rutten T, Lloyd CW (1996) Isolation of microtubule-associated proteins from carrot cytoskeletons: a 120 kDa map decorates all four microtubule arrays and the nucleus. Plant J 10: 251-259

Chan MM, Fong D (1990) Inhibition of leishmanias but not host macrophages by the antitubulin herbicide trifluralin. Science 249: 924-926

Chan MM, Fong D (1994) Plant microtubule inhibitors against trypanosomatids. Parasitol Today 10: 448-451

Chan MM, Grogl M, Bienen EJ, Fong D (1993) Herbicides to curb human parasitic infections: in vitro and in vivo effects of trifluralin on the trypanosomatid protozoans. Proc Natl Acad Sci USA 90: 5657-5661

Chang-Jie J, Sonobe S (1993) Identification and preliminary characterization of a 65 kD higher-plant microtubule-associated protein. J Cell Sci 105: 891-901

Chu B, Wilson TJ, McCune-Zierath C, Snustad DP, Carter JV (1998) Two β–tubulin genes, *TUB1* and *TUB8*, of *Arabidopsis* exhibit largely nonoverlapping patterns of expression. Plant Mol Biol 37: 785-790

Cleary AL, Smith LG (1998) The *Tangled1* gene is required for spatial control of cytoskeletal arrays associated with cell division during maize leaf development. Plant Cell 10: 1875-1888

Cleveland DW (1987) The multitubulin hypothesis revisited: what have we learned? J Cell Biol 104: 381-383

Cruz MC, Edlind T (1997) Beta–Tubulin genes and the basis for benzimidazole sensitivity of the opportunistic fungus *Cryptococcus neoformans*. Microbiology 143: 2003-2008

Cyr RJ (1991) Microtubule-associated proteins in higher plants. In: Lloyd CW (ed) The Cytoskeletal of Plant Growth and Form. Academic Press, New York, pp 57-67

Cyr RJ (1994) Microtubules in plant morphogenesis: role of the cortical array. Annu Rev Cell Biol 10: 153-180

Davidse LC (1986) Benzimidazole fungicides: mechanism of action and biological impact. Ann Rev Phytopathol 24: 43-65

Davidse LC, Flach W (1977) Differential binding of methylbenzimidazole-2-yl carbamate to fungal tubulin as a mechanism of resistance to this antimitotic agent in mutant strains of *Aspergillus nidulans*. J Cell Biol 72: 174-193

Delp CJ (1980) Coping with resistance to plant disease control agents. Plant Dis 64: 652-657

Durso NA, Cyr RJ (1994) A calmodulin-sensitive interaction between microtubules and a higher plant homolog of elongation factor 1α. Plant Cell 6: 893-905

Ebneth A, Godemann R, Stramer K, Illenberger S, Trinczek B, Mandelkow EM, Mandelkow E (1998) Overexpression of tau protein inhibits kinesin-dependent trafficking of vesicles, mitochondria, and endoplasmic reticulum: Implications for Alzheimer's disease. J Cell Biol 143: 777-794

Ellis JR, Taylor R, Hussey PJ (1994) Molecular modeling indicates that two chemically distinct classes of anti-mitotic herbicides bind to the same receptor site(s). Plant Physiol 105: 15-18

Erickson HP, O'Brien ET (1992) Microtubule dynamic instability and GTP hydrolysis. Annu Rev Biophys Biomol Struct 21: 145-166

Fisher RH, Barton MK, Cohen JD, Cooke TJ (1996) Hormonal studies of *fass*, an *Arabidopsis* mutant that is altered in organ elongation. Plant Physiol 110: 1109-1121

Fosket DE (1989) Cytoskeletal proteins and their genes in higher plants. In: Stumpf PK, Conn, EE (eds) The biochemistry of plants. Academic Press, San Diego, pp 393-454.

Fosket DE, Morejohn LC (1992) Structural and functional organization of tubulin. Annu Rev Plant Physiol Plant Mol Biol 43: 201-240

Foster KE, Burland TG, Gull K (1987) A mutant beta–tubulin confers resistance to the action of benzimidazole-carbamate microtubule inhibitors both in vivo and in vitro. Eur J Biochem 163: 449-455

Francis D, Dudits D, Inze D (1998) Plant cell division. Portland Press, London, pp 347

Fujimura M, Oeda K, Inoue H, Kato T (1990) Mechanism of action of N-phenylcarbamates in benz-imidazole-resistant *Neurospora* strains. In: Green MB, LeBaron HM, Moberg WK (eds) Managing resistance to agrochemicals. American Chemical Society, Washington, DC, pp 224-236

Fujimura M, Kamakura T, Inoue H, Inoue S, Yamaguchi I (1992a) Sensitivity of *Neurospora crassa* to benzimidazoles and N-phenylcarbamates: effect of amino acid substitutions at position 198 in beta–tubulin. Pestic Biochem Physiol 44: 165-173

Fujimura M, Oeda K, Inoue H, Kato T (1992b) A single amino-acid substitution in the β–tubulin gene of *Neurospora crassa* confers both carbendazim resistance and diethofencarb sensitivity. Curr Genet 21: 399-404

Goddard RH, Wick SM, Silflow CD, Snustad DP (1994) Microtubule components of the plant cyto-skeleton. Plant Physiol 104: 1-6

Gossett BJ, Murdock EC, Toler JE (1992) Resistance of palmer amaranth (*Amaranthus palmeri*) to the dinitroaniline herbicides. Weed Tech 6: 587-591

Gu X, Verma DPS (1996) Phragmoplastin, a dynamin-like protein associated with cell plate formation in plants. EMBO J 15: 695-704

Gu X, Verma DPS (1997) Dynamics of phragmoplastin in living cells during cell plate formation and uncoupling of cell elongation from the plane of cell division. Plant Cell 9: 157-169

Gunning BES (1982) The cytokinetic apparatus: its development and spatial regulation. In: Lloyd CW (ed) The cytoskeleton in plant growth and development. Academic Press, New York, p 230-288

Hardtke CS, Berleth T (1998) The *Arabidopsis* gene *MONOPTEROS* encodes a transcription factor mediating embryo axis formation and vascular development. EMBO J 17: 1405-1411

Heap IM (1997) The occurrence of herbicide-resistant weeds worldwide. Pestic Sci 51: 235-243

Hepler PK, Bonsignore CL (1990) Caffeine inhibition of cytokinesis: ultrastructure of cell plate forma-tion/degredation. Protoplasma 157: 182-192

Hepler PK, Hush JM (1996) Behavior of microtubules in living plant cells. Plant Physiol 112: 455-461

Hess FD, Bayer DE (1977) Binding of the herbicide trifluralin to *Chlamydomonas* flagellar tubulin. J Cell Sci 24: 351-360

Hightower RC, Meagher RB (1986) The molecular evolution of actin. Genetics 114: 315-332

Hiraoka Y, Toda T, Yanagida M (1984) The *NDA3* gene of fission yeast encodes β–tubulin: a cold sensitive *nda3* mutation reversibly blocks spindle formation and chromosome movement in mitosis. Cell 39: 349-358

Hirata D, Masuda H, Eddison M, Toda T (1998) Essential role of tubulin-folding cofactor D in micro-tubule assembly and its association with microtubules in fission yeast. EMBO J 17: 656-666

Hoffman JC, Vaughn KC (1994) Mitotic disrupter herbicides act by a single mechanism but vary in efficacy. Protoplasma 179: 16-25

Hoffman JC, Vaughn KC (1995a) Post-translational tubulin modifications in spermatogeneous cells of the pteridophyte *Ceratopteris richardii*. Protoplasma 186: 169-182

Hoffman JC, Vaughn KC (1995b) Using the developing spermatogenous cells of *Ceratopteris* to unlock the mysteries of the plant cytoskeleton. Int J Plant Sci 156: 346-358

Huffaker TC, Hoyt MA, Botstein D (1987) Genetic analysis of the yeast cytoskeleton. Annu Rev Genet 21: 259-284

Hugdahl JD, Morejohn LC (1993) Rapid and reversible high-affinity binding of the dinitroaniline herbicide oryzalin to tubulin from *Zea mays* L. Plant Physiol 102: 725-740

Hugdahl JD, Bokros CL, Hanesworth VR, Aalund GR, Morejohn LC (1993) Unique functional char-acteristics of the polymerization and MAP binding regulatory domains of plant tubulin. Plant Cell 5: 1063-1080

Hugdahl JD, Bokros CL, Morejohn LC (1995) End-to-end annealing of plant microtubules by the p86 subunit of eukaryotic initiation factor-(iso)4F. Plant Cell 7: 2129-2138

Hush JM, Wadsworth P, Callaham DA, Hepler PK (1994) Quantification of microtubule dynamics in living plant cell using fluorescence redistribution after photobleaching. J Cell Sci 107: 775-784

Hussey PJ, Snustad DP, Silflow CD (1991) Tubulin gene expression in higher plants. In: Lloyd CW (ed) The cytoskeletal basis of plant growth and form. Academic Press, New York, pp 15-27

James EH, Kemp MS, Moss SR (1995) Phytotoxicity of trifluoromethyl- and methyl-substituted dini-troaniline herbidices on resistant and susceptible populations of black-grass (*Alopecurus myosuroides*). Pestic Sci 43: 273-277

James SW, Lefebvre PA (1992) Genetic interactions among *Chlamydomonas reinhardtii* mutations that confer resistance to anti-microtubule herbicides. Genetics 130: 305-314

James SW, Silflow CD, Lefebvre PA (1988) Mutants resistant to anti-microtubule herbicides map to a locus on the uni linkage group in *Chlamydomonas reinhardtii*. Genetics 118: 141-147

James SW, Silflow CD, Stroom P, Lefebvre PA (1993) A mutation in the α1-tubulin gene of *Chlamydomonas reinhardtii* confers resistance to anti-microtubule herbicides. J Cell Sci 106: 209-218

Jasieniuk M, Brule-Babel AL, Morrison IN (1994) Inheritance of trifluralin resistance in green foxtail (*Setaria viridis*). Weed Sci 42: 123-127

Jones AL, Shabi E, Ehret G (1987) Genetics of negatively correlated cross-resistance to a N-phenylcarbamate in benomyl-resistant *Venturia inaequalis*. Can J Plant Pathol 9: 195-199

Joyce CM, Villemur R, Snustad DP, Silflow CD (1992) Tubulin gene expression in maize (*Zea mays* L.): change in isotype expression along the developmental axis of seedling roots. J Mol Biol 227: 97-107

Jung MK, Oakley BR (1990) Identification of an amino acid substitution in the benA, β–tubulin gene of *Aspergillus nidulans* that confers thiabendazole resistance and benomyl supersensitivity. Cell Motil Cytoskel 17: 87-94

Jung MK, Wilder IB, Oakley BR (1992) Amino acid alterations in the benA (ß-tubulin) gene of *Aspergillus nidulans* that confer benomyl resistance. Cell Motil Cytoskeleton 22: 170-174

Jürgens G, Mayer U, Torres-Ruiz RA, Berleth T, Miséra S (1991) Genetic analysis of pattern formation in the *Arabidopsis* embryo. Development Suppl 91: 27-38

Kamada T, Sumiyoshi T, Takemaru T (1989) Mutations in β–tubulin block transhyphal migration of nuclei in dikaryosis in the homoblasidiomycete *Coprinus cinereus*. Plant Cell Physiol 30: 1073-1080

Kamada T, Hirami H, Sumiyoshi T, Tanabe S, Takemaru T (1990) Extragenic suppressor mutations of a β–tubulin mutation in the basidiomycete *Coprinus cinereus*: isolation and genetic and biochemical analyses. Curr Microbiol 20: 223-228

Katz W, Weinstein B, Solomon F (1990) Regulation of tubulin levels and microtubule assembly in *Saccharomyces cerevisiae*: consequences of altered tubulin gene copy number in yeast. Mol Cell Biol 10: 2730-2736

Kilmartin J (1981) Purification of yeast tubulin by self-assembly in vitro. Biochemistry 20: 3629-3633

Kirschner M, Mitchison T (1986) Beyond self-assembly: from microtubules to morphogenesis. Cell 45: 329-342

Koenraadt H, Jones AL (1993) Resistance to benomyl conferred by mutations in codon 198 or 200 of the beta–tubulin gene of *Neurospora crassa* and sensitivity to diethofencarb conferred by codon 198. Phytopathology 83: 850-854

Koenraadt H, Somerville SC, Jones AL (1992) Characterization of mutations in the beta–tubulin gene of benomyl-resistant field strains of *Venturia inaequalis* and other plant pathogenic fungi. Phytopathology 82: 1348-1354

Kopczak SD, Haas NA, Hussey PJ, Silflow CD, Snustad DP (1992) The small genome of *Arabidopsis* contains at least six expressed α–tubulin genes. Plant Cell 4: 539-547

Kozminski KG, Diener DR, Rosenbaum JL (1993) High level expression of nonacetylatable α–tubulin in *Chlamydomonas reinhardtii*. Cell Motil Cytoskel 25: 158-170

Lambert AM (1993) Microtubule-organizing centers in higher plants. Curr Opin Cell Biol 5: 116-122

Laporte K, Rossignol M, Trass JA (1993) Interaction of tubulin with the plama membrane: tubulin is present in purified plasmalemma and behaves as an integral membrane protein. Planta 191: 413-416

Lauber MH, Waizenegger I, Steinmann T, Schwarz H, Mayer U, Hwang I, Ludkowitz W, Jürgens G (1997) The *Arabidopsis KNOLLE* protein is a cytokinesis-specific syntaxin. J Cell Biol 139: 1485-1493

Lechtreck KF, Melkonian M (1991) An update on fibrous flagellar roots in green algae. Protoplasma 164: 38-44

Lee VD, Huang B (1990) Missense mutations at lysine 350 in ß2-tubulin confer altered sensitivity to microtubule inhibitors in *Chlamydomonas*. Plant Cell 2: 1051-1057

Lewis SA, Cowan NJ (1988) Complex regulation and functional versatility of mammalian alpha and beta–tubulin isotypes during differentiation of testis and muscle cells. J Cell Biol 106: 2023-2033

Li J, Katiyar SK, Edlind TD (1996) Site-directed mutagenesis of *Saccharomyces cerevisiae* ß-tubulin: interaction between residue 167 and benzimidazole compounds. FEBS Lett 385: 7-10

Liaud MF, Brinkmann H, Cerff R (1992) The ß-tubulin gene family of pea: primary structures, genomic organization and intron-dependent evolution of genes. Plant Mol Biol 18: 639-651

Little M, Ludueña RF (1985) Structural differences between brain β1- and β2-tubulin: implications for microtubule assembly and colchicine binding. EMBO J 4: 51-56

Liu B, Marc J, Joshi HC, Palevitz BA (1993) A γ–tubulin-related protein associated with the microtubule arrays of higher plants in a cell cycle-dependent manner. J Cell Sci 104: 1217-1228

Liu B, Joshi H, Wilson TJ, Silflow CD, Palevitz BA, Snustad DP (1994) γ–Tubulin in *Arabidopsis:* gene sequence, immunoblot, and immunofluorescence studies. Plant Cell 6: 303-314

Liu B, Cyr RJ, Palevitz BA (1996) A kinesin-like protein, KatAp, in the cells of *Arabidopsis* and other plants. Plant Cell 8: 119-132

Liu CM, Meinke DW (1998) The titan mutants of *Arabidopsis* are disrupted in mitosis and cell cycle control during seed development. Plant J 16: 21-31

Liu CM, Johnson S, Wang TL (1995) cyd, A mutant of pea that alters embryo morphology is defective in cytokinesis. Dev Genet 16: 321-331

Lloyd CW (1987) The Plant Cytoskeleton: The impact of fluorescence microscopy. Annu Rev Plant Physiol 38: 119-139

Lloyd CW (1991) The cytoskeletal basis of plant growth and form. Academic Press, London.

Lopez I, Khan S, Sevik M, Cande WZ, Hussey PJ (1995) Isolation of a full length cDNA encoding *Zea mays* gamma–tubulin. Plant Physiol 107: 309-310

Ludueña RF (1998) Multiple forms of tubulin: different gene products and covalent modifications. Int Rev Cytol 178: 207-275

Lukowitz W, Mayer U, Jürgens G (1996) Cytokinesis in the *Arabidopsis* embryo involves the syntaxin-related *KNOLLE* gene product. Cell 84: 61-71

Machin NA, Lee JM, Barnes G (1995) Microtubule stability in budding yeast: characterization and dosage suppression of a benomyl-dependent tubulin mutant. Mol Biol Cell 6: 1241-59

MacRae TH (1997) Tubulin post-translation modifications: enzymes and their mechanisms of action. Eur J Biochem 244: 265-278

Marc J (1997) Microtubule-organizing centres in plants. Trends Plant Sci 2: 223-230

Marc J, Sharkey DE, Durso NA, Zhang M, Cyr RJ (1996) Isolation of a 90-kD microtubule-associated protein from Tobacco membranes. Plant Cell 8: 2127-2138

Marschall LG, Jeng RL, Mulholland J, Stearns T (1996) Analysis of Tub4p, a yeast γ–tubulin-like protein: implications for microtubule-organizing center function. J Cell Biol 134: 443-454

May GS, Tsang MLS, Smith H, Fidel S, Morris NR (1987) *Aspergillus nidulans* β–tubulin genes are unusually divergent. Gene 55: 231-243

Mayer U, Ruiz RAT, Berleth T, Mísera S, Jürgens G (1991) Mutations affecting body organization in the *Arabidopsis* embryo. Nature 353: 402-407

Mayer U, Buttner G, Jürgens G (1993) Apical-basal pattern formation in the *Arabidopsis* embryo: studies on the role of the *gnom* gene. Development 117: 149-162

Mazars TL, Nacry C, Bouchez P, Moreau M, Ranjeva R, Thuleau P (1998) Plasma membrane depolarization- activated calcium channels, stimulated by microtubule- depolymerizing drugs in wild-type *Arabidopsis thaliana* protoplasts, display constitutively large activities and a longer half-life in *ton 2* mutant cells affected in the organization of cortical microtubules. Plant J 13: 603-610

McAlister FM, Holtum JAM, Powles SB (1995) Dinitroaniline herbicide resistance in rigid ryegrass (*Lolium rigidum*). Weed Sci 43: 55-62

McClinton RS, Sung ZR (1997) Organization of cortical microtubules at the plasma membrane in *Arabidopsis*. Planta 201: 252-260

McIntosh K, Pickett-Heaps JD, Gunning BES (1995) Cytokinesis in *Spirogyra*: integration of cleavage and cell-plate formation. Int J Plant Sci 156: 1-8

McKay GJ, Cooke LR (1997) A PCR-based method to characterize and identify benzimidazole resistance in *Helminthosporium solani*. FEMS Microbiol Lett 152: 371-378

Meagher RB, Williamson RE (1994) The Plant Cytoskeleton. In: Meyerowitz E, Sommerville C (eds) *Arabidopsis*. Cold Spring Harbor Laboratory Press, Cold Spring Harbor, pp 1049-1084

Meinke DW (1986) Embryo-lethal mutants and the study of plant embryo development. Oxf Surv Plant Mol Cell Biol 3: 122-165

Meinke DW (1991) Perspectives on genetic analysis of plant embryogenesis. Plant Cell 3: 857-866

Morejohn LC, Bureau T, Mole-Bajer J, Bajer AS, Fosket DE (1987) Oryzalin, a dinitroaniline herbicide, binds to plant tubulin and inhibits microtubule polymerization in vitro. Planta 172: 252-264

Morris NR (1986) The molecular gentics of of microtubule proteins in fungi. Exp Mycol 10: 77-82

Morris NR, Lai MH, Oakley CE (1979) Identification of a gene for α–tubulin in *Aspergillus nidulans*. Cell 16: 437-442

Morrison IN, Todd BG, Nawolsky KM (1989) Confirmation of trifluralin-resistant green foxtail (*Setaria viridis*) in Manitoba. Weed Technol 3: 544-551

Moss SR (1990) Herbicide cross-resistance in slender foxtail (*Alopecurus myosuroides*). Weed Sci 38: 492-496

Mudge LC, Gossett BJ, Murphy TR (1984) Resistance of goosegrass (*Eleusine indica*) to dinitroaniline herbicides. Weed Sci 32: 591-594

Murphy TR, Gossett BJ, Toler JE (1986) Growth and development of dinitroaniline-susceptible and resistant goosegrass (*Eleusine indica*) biotypes under noncompetitive conditions. Weed Sci 34: 704-710

Murthy JV, Kim HH, Hanesworth VR, Hugdahl JD, Morejohn LC (1994) Competitive inhibition of high affinity oryzalin binding to plant tubulin by the phosphoric amide herbicide amiprophos-methyl. Plant Physiol 105: 309-319

Mysore K, Baird V (1995) Molecular characterization of the tubulin-related gene families in herbicide resistant and susceptible goosegrass (*Eleusine indica*). Weed Sci 43: 28-33

Neff NF, Thomas JH, Grisafi P, Botstein D (1983) Isolation of the β–tubulin gene from yeast and demonstration of its essentail function in vivo. Cell 33: 211-219

Nikolova G, Baeva G (1996) Pendimethalin-resistant biotypes of *Echinochola crus-galli* found in Bulgaria. Int Symp on Weed and Crop Resistance to Herbicides, Cordoba, Spain, April 3-6, 1995, pp 32-33

Nogales E, Wolf SG, Downing KH (1998) Structure of the αβ tubulin dimer by electron crystallography. Nature 391: 199-202

Nogales E, Whittaker M, Milligan RA, Downing KH (1999) High-resolution model of the microtubule. Cell 96: 79-88

Oakley BR (1985) Microtubule mutants. Can J Biochem Cell Biol 63: 479-488

Oakley BR (1992) γ–Tubulin: the microtubule organizer? Trends Cell Biol 2: 1-5

Oakley BR (1994) γ–Tubulin. In: Hyams JS, Lloyd CW (eds) Microtubules. Wiley-Liss, New York, pp 33-45

Oakley BR (1999) Methods for isolating and analyzing mitotic mutants in *Aspergillus nidulans*. In: Rieder CL (ed) Methods in Cell Biology. Academic Press, London, pp 347-368

Oakley BR, Morris NR (1980) Nuclear movement is β–tubulin dependent in *Aspergillus nidulans*. Cell 19: 255-262

Oakley BR, Morris NR (1981) A β–tubulin mutation in *Aspergillus nidulans* that blocks microtubule function without blocking assembly. Cell 24: 837-845

Oakley BR, Oakley CE, Rinehart JE (1987) Conditionally lethal *tubA* α–tubulin mutations in *Aspergillus nidulans*. Mol Gen Genet 208: 135-144

Oakley BR, Oakley CE, Yoon Y, Jung MK (1990) γ–Tubulin is a component of the spindle pole body that is essential for microtubule function in *Aspergillus nidulans*. Cell 61: 1289-1301

Oakley CE, Oakley BR (1989) Identification of γ–tubulin, a new member of the tubulin superfamily encoded by *mipA* gene of *Aspergillus nidulans*. Nature 338: 662-664

Orbach MJ, Porro EB, Yanofsky C (1986) Cloning and characterization of the gene for ß-tubulin from a benomyl-resistant mutant of *Neurospora crassa* and its use as a dominant selectable marker. Mol Cell Biol 6: 2452-2461

Qin X, Gianì S, Breviario D (1997) Molecular cloning of three rice α–tubulin isotypes: differential expression in tissues and during flower development. Biochim et Biophys Acta 1354: 19-23

Raghavan V (1997) Molecular embryology of flowering plants. Cambridge University Press, Cambridge and New York.

Raper JR (1966) Genetics of sexuality in higher fungi. Ronald, New York, p. 283

Reijo RA, Cooper EM, Beagle GJ, Huffaker TC (1994) Systematic mutational analysis of the yeast beta–tubulin gene. Mol Biol Cell 5: 29-43

Rogers HJ, Greenland AJ, Hussey PJ (1993) Four members of the maize beta–tubulin gene family are expressed in the male gametophyte. Plant J 4: 875-882

Röper W, Röper S (1977) Centripetal wall formation in roots of *Vicia faba* after caffeine treatment. Protoplasma 93: 89-100

Rutten T, Chan J, Lloyd CW (1997) A 60-kDa plant microtubule-associated protein promotes the growth and stabilization of neurotubules in vitro. Proc Natl Acad Sci USA 94: 4469-4474

Sage CR, Dougherty CA, Sullivan K, Farrell KW (1995) β–Tubulin mutation suppresses microtubule dynamics in vitro and slows mitosis in vivo. Cell Motil Cytoskel 30: 285-300

Schatz PJ, Solomon F, Botstein D (1988) Isolation and characterization of the conditional-lethal mutation in the *TUB1* α–tubulin gene of the yeast *Saccharomyces cerevisiae*. Genetics 120: 681-695

Schibler MJ, Huang B (1991) The *colR4* and *colR15* ß-tubulin mutations in *Chlamydomonas* confer altered sensitivities to microtubule inhibitors and herbicides by enhancing microtubule stability. J Cell Biol 113: 605-614

Schmit AC, Stoppin V, Chevrier V, Job D, Lambert AM (1994) Cell cycle dependent distribution of a centrosomal antigen at the perinuclear MTOC or at the kinetochores of higher plant cells. Chromosoma 103: 343-351

Seagull RW (1989) The plant cytoskeleton. Crit Rev Plant Sci 8: 131-167

Sheir-Neiss G, Lai MH, Morris NR (1978) Identification of a gene for β–tubulin in *Aspergillus nidulans*. Cell 15: 639-647

Sheridan WF (1995) Genes and embryo morphogenesis in angiosperms. Dev Gen 16: 291-297

Sheridan WF, Clark JK (1993) Mutational analysis of morphogenesis of the maize embryo. Plant J 3: 347-358

Shevell DE, Leu WM, Gillmor CS, Xia G, Feldmann KA, Chua NH (1994) *EMB30* is essential for normal cell division, cell expansion, and cell adhesion in *Arabidopsis* and encodes a protein that has similarity to Sec7. Cell 77: 1051-1062

Shibaoka H, Nagai R (1994) The plant cytoskeleton. Curr Opin Cell Biol 6: 10-15

Silflow CD, Oppenheimer DG, Kopczak SD, Ploense SE, Ludwig SR, Haas N, Snustad DP (1987) Plant tubulin genes: structure and differential expression during development. Dev Genet 8: 435-460

Smeda RJ Vaughn KC (1994) Resistance to dinitroaniline herbicides. In: Powels SB, Holtum JAM (eds) Herbicide resistance in plants: Biology and Biochemistry. CRC Press, Boca Raton, pp 215-228

Smertenko A, Blume YB, Viklicky V, Dráber P (1997a) Exposure of tubulin structural domains in *Nicotiana tabacum* microtubules probed by monoclonal antibodies. Eur J Cell Biol 72: 104-112

Smertenko A, Blume YB, Viklicky V, Opatrný Z, Dráber P (1997b) Post-translational modifications and multiple isoforms of tubulin in *Nicotiana tabacum* cells. Planta 201: 349-358

Smith AM, Archer JE, Solomon F (1998) Reguation of tubulin polypeptides and microtubule function: Rki1p interacts with the β–tubulin binding protein Rbl2p. Chromosoma 107: 471-478

Smith LG, Hake S, Sylvester AW (1996) The *tangled-1* mutation alters cell division orientations throughout maize leaf development without altering leaf shape. Development 122: 481-489

Snustad DP, Haas NA, Kopczak SD, Silflow CD (1992) The small genome of *Arabidopsis thaliana* contains at least nine expressed ß-tubulin genes. Plant Cell 4: 549-556

Sobel SG, Snyder M (1995) A highly divergent γ–tubulin gene is essential for cell growth and proper microtubule organization in *Saccharomyces cerevisiae*. J Cell Biol 131: 1775-1788

Stearns T, Botstein D (1988) Unlinked noncomplementation: isolation of new conditional-lethal mutations in each of the tubulin genes of *Saccharomyces cerevisiae*. Genetics 119: 249-260

Stephens RE (1995) Ciliary membrane tubulin: isolation and fractionation. Methods Cell Biol 47: 431-436

Stoppin V, Lambert AM, Vantard M (1996) Plant microtubule-associated proteins (MAPs) affect microtubule nucleation and growth at plant nuclei and mammalian centrosomes. Eur J Cell Biol 69: 211-230

Thomas JH, Neff NF, Botstein D (1985) Isolation and characterization of mutations in the ß-tubulin gene of *Saccharomyces cerevisiae*. Genetics 111: 715-734

Toda T, Umesono K, Hirata A, Yanagida M (1983) Cold-sensitive nuclear division arrests mutants of the fission yeast *Schizosaccharomyces pombe*. J Mol Biol 168: 251-270

Toda T, Adachi Y, Hiraoka Y, Yanagida M (1984) Identification fo the pleiotropic cell division cycle gene *NDA2* as one of two different α–tubulin genes in *Schizosaccharomyces pombe*. Cell 37: 233-242

Torres-Ruiz RA, Jürgens G (1994) Mutations in the *FASS* gene uncouple pattern formation and morphogenesis in *Arabidopsis* development. Development 120: 2967-2978

Traas J, Laufs P (1998) Cell cycle mutants in higher plants: a phenotypical overview. In: Francis D, Dudits D, Inze D (eds) Plant Cell Division. Portland Press, London, pp 319-336

Traas J, Bellini C, Nacry P, Kronenberger J, Bouchez D, Caboche M (1995) Normal differentiation patterns in plants lacking microtubular preprophase bands. Nature 375: 676-677

Ulmasov T, Hagen G, Guilfoyle TJ (1997) ARF1, a transcription factor that binds to auxin response elements. Science 276: 1865-1868

Umesono K, Toda T, Hayashi S, Yanagida M (1983) Two cell division cycle genes *NDA2* and *NDA3* of the fission yeast *Schizosaccharomyces pombe* control microtubular organization and sensitivity to anti-mitotic benzimidazole compounds. J Mol Biol 168: 271-284

Uribe X, Torres MA, Capellades M, Puigdomènech P, Rigau J 1998. Maize α–tubulin genes are expressed according to specific patterns of cell differentiation. Plant Mol Biol 37: 1069-1078

Vaughan MA, Vaughn KC (1987) Taxol treatment of *Eleusine* indicates hyper-stabilized tubulin may cause dinitroaniline resistance. Plant Physiol (Suppl) 83: 107

Vaughan MA, Vaughn KC (1988) Carrot microtubules are dinotroaniline resistant. I. Cytological and cross-resistance studies. Weed Res 28: 73-83

Vaughn KC, Harper JDI (1998) Microtubule-organizing centers and nucleating sites in land plants. Int Rev Cytol 181: 75-149

Vaughn KC, Koskinen WC (1987) Effects of trifluralin metabolites on goosegrass (*Eleusine indica*) root meristems. Weed Sci 35: 36-42

Vaughn KC, Vaughan MA (1986) Dinitroaniline resistance in *Eleusine* is due to altered tubulin. Plant Physiol (Suppl) 80: 67

Vaughn KC, Vaughan MA (1990) Structural and biochemical characterization of dinitroaniline-resistant *Eleusine*. In: Green MB, LeBaron HM, Moberg WK (eds) Managing resistance to agrochemicals: from fundamental research to practical strategies. ACS Symp Series. American Chemical Society, Los Angeles, pp 364-375

Vaughn KC, Marks MD, Weeks DP (1987) A dinitroaniline-resistant mutant of *Eleusine indica* exhibits cross-resistance and susceptibility to antimicrotubule herbicides and drugs. Plant Physiol 83: 956-964

Vaughn KC, Vaughan MA, Gossett BJ (1990) A biotype of goosegrass (*Eleusine indica*) with an intermediate level of dinitroaniline resistance. Weed Technol 4: 157-162

Vega LR, Fleming J, Solomon F (1998) An α–tubulin mutant destabilizes the heterodimer: phenotypic consequences and interactions with tubulin-binding proteins. Mol Biol Cell 9: 2349-2360

Villemur R, Joyce CM, Haas NA, Goddard RH, Kopczak SD, Hussey PD, Snustad DP, Silflow CD (1992) α–Tubulin gene family of maize (*Zea mays* L.). Evidence for two ancient α–tubulin genes in plants. J Mol Biol 227: 81-96

Villemur R, Haas NA, Joyce CM, Snustad DP, Silflow CD (1994 Characterization of four new ß-tubulin genes and their expression during male flower development in maize (*Zea mays* L.). Plant Mol Biol 24: 295-315

Waldin T, Ellis R, Hussey P (1992) Tubulin-isotype analysis of two grass species resistant to dinitro-aniline herbicides. Planta 188: 258-264

Waldin TR (1995) Analysis of dinitroaniline and phosphorothioamidate resistant grasses. PhD Thesis, University of London, London

Wang T, Fleury A, Ma J, Darmency H (1996) Genetic control of dinitroaniline resistance in foxtail millet (*Setaria italica*). J Hered 87: 423-426

Wehland J, Schroeder M, Weber K (1984) Organization of microtubules in stabilized meristematic plant cells revealed by a rat monoclonal antibody reacting only with the tyrosinated form of al-pha–tubulin. Cell Biol Int Rep 8: 147-150

Weinstein B, Solomon F (1990) Phenotypic consequences of tubulin overproduction in *Saccharomyces cerevisiae*: differences between alpha–tubulin and beta–tubulin. Mol Cell Biol 10: 5295-5304

Whittaker DJ, Triplett BA (1999) Gene-specific changes in α–tubulin transcript accumulation in developing cotton fibers. Plant Physiol 121: 181-188

Wiche G, Oberkanins C, Himmler A (1991) Molecular structure and function of microtubule-associated proteins. Int Rev Cytol 124: 217-273

Wymer CL, Fisher DD, Moore RC, Cyr RJ (1996) Elucidating the mechanism of cortical microtubule reorientation in plant cells. Cell Motil Cytoskelet 35: 162-173

Yamamoto E, Baird WV (1999) Molecular characterization of four ß-tubulin genes from dinitroaniline susceptible and resistant biotypes of *Eleusine indica*. Plant Mol Biol 39: 45-61

Yamamoto E, Zeng L, Baird WV (1998) α–tubulin missense mutations correlate with antimicrotubule drug resistance in *Eleusine indica*. Plant Cell 10: 297-308

Yelverton F, Isgrigg I (1998) Herbicide-resistant weeds in turfgrass. Golf Course Management December 66: 56-61

Yemets AI, Blume YB (1999) Resistance to herbicides with antimicrotubular activity: From natural mutants to transgenic plants. Russ J Plant Physiol 46: 786-796

Yemets AI, Kundel'chuk OP, Smertenko AP, Solodushko VG, Rudas VA, Gleba YY, Blume YB (2000) Transfer of amiprophosmethyl-resistance from *Nicotiana plumbaginifolia* by somatic hybridization. Theor Appl Genet (in press)

Yemets AI, Smertenko AP, Solodushko VG, Kundel'chuk OP, Blume YB (1997) Asymmetric somatic hybrids with mutant ß-tubulin resistant to amiprophosmethyl. In: Tewari KK, Singhal GS (eds) Plant molecular biology and biotechnology. Narosa Publishing House, New Delhi, India, pp 220-234

Yemets AI, Strashnyuk NM, Blume YB (1997b) Plant mutants and somatic hybrids with resistance to trifluralin. Cell Biol Int 21: 912-914

Yuan M, Shaw PJ, Warn RM, Lloyd CW (1994) Dynamic reorientation of cortical microtubules, from transverse to longitudinal, in living plant cells. Proc Natl Acad Sci USA 91: 6050-6053

Zeng L, Baird WV (1997) Genetic basis of dinitroaniline herbicide resistance in a highly-resistant biotype of goosegrass (*Eleusine indica*). J Hered 88: 427-432

Zeng L, Baird WV (1999) Inheritance of resistance to anti-microtubule dinitroaniline herbicides in an Intermediate resistant biotype of *Eleusine indica* (Poaceae). Am J Bot 86: 940-947

Zhang DH, Wadsworth P, Hepler PK (1990) Microtubule dynamics in living plant cells: confocal imaging of microinjected fluorescent brain tubulin. Proc Natl Acad Sci USA 87: 8820-8824

Zhang DH, Wadsworth P, Hepler PK (1993) Dynamics of microfilaments are similar but distinct from microtubules during cytokinesis in living, dividing plant cells. Cell Motil Cytoskelet 24: 151-155

9 Anticytoskeletal Herbicides

Kevin C. Vaughn
USDA-ARS, SWSL, P.O. Box 350, Stoneville, Minnesota 38776-0350, USA

9.1
Summary

Approximately one quarter of all of the marketed herbicides are classified into the mitotic disrupter herbicide group, including the widely used dinitroaniline and carbamate herbicides. Most of the herbicides in this group are used commercially to control grasses and other small-seeded weed species in larger-seeded dicot crop species. Gross morphology of the seedlings after treatment with mitotic disrupter herbicides is distinctly club-shabed or swollen, compared to the uniformly tapered roots found in untreated controls, and is similar to those effects noted for the classical microtubule disrupter colchicine. Microscopic examination of herbicide-treated roots reveals a concentration-dependent loss of microtubules, with phragmoplast and spindle arrays being affected at lowest concentration, and cortical and kinetochore microtubules being the least affected. The loss of these microtubule arrays results in the production of irregular cell walls, C-metaphase figures and lobed nuclei. Loss of the cortical microtubule array results in isodiametric growth, which leads to root clubbing in the zone of root elongation. Limited biochemical analysis indicates that the herbicides oryzalin and pronamide (propyzamide) bind directly to tubulin and that the carbamates and phosphoric amides can inhibit polymerization in vitro, indirectly confirming this mode of action for these herbicides as well. A possible non-tubulin target has been suggested for the herbicide dithiopyr, although the effects induced are identical to other members of this group. Data from both resistant mutants and molecular modelling indicate that the dinitroaniline and phosphoric amides bind to a similar site, possibly on the alpha-tubulin molecule. Microtubule disrupter herbicides, because of their high selectivity, have great potential in screens for mutants with altered herbicide sensitivity that could be of tremendous agronomic importance.

9.2
Agronomic aspects of mitotic disrupter herbicides

The mitotic disrupter or microtubule disrupter group of herbicides (Vaughn and Lehnen 1991; Molin and Khan 1997) accounts for about a quarter of all the herbicides marketed. Herbicides in this group include the dinitroanilines, such as triflu-ralin and oryzalin, the phosphoric amides, such as amiprophos methyl, and the carbamate herbicides, such as chlorpropham (CIPC) and propham (IPC). Other widely used mitotic disrupter herbicides include DCPA, dithiopyr and pronamide (propyzamide). Examples of some of the chemical structures of the compounds in this group are included in Fig. 9.1.

Although chemically variable as a group, these herbicides do have much in common in their uses and symptomatology. As a group, they tend to be very lipophilic. Most are used as pre-emergent herbicides, often by incorporating the herbicide into the soil (Hess 1987; Molin and Khan 1997). Because the movement of these herbicides is often limited, the incorporation of the herbicide into a shallow depth of soil will arrest the development of the seedling of small-seeded species so that the seedling dies before reaching a zone of soil free of herbicides. These small seedlings often have club-shaped roots after herbicide treatment (e.g. Vaughan and Vaughn 1988), which is one of the easy visual tests for herbicides with this mode of action, although cellulose biosynthesis inhibitor herbicides cause a similar effect (Sabba and Vaughn 2000). Larger-seeded weeds and crops will have enough nutrient resources to grow through the zone of incorporation, although there will be an area of root pruning (a zone with few secondary roots) that marks the zone of herbicide incorporation in the soil (Hess 1987). Once deeper than the zone of incorporation, normal root growth resumes. In general, dicots are much less sensitive to these compounds than are monocots, perhaps due to the higher quantities of lipids in dicot seedlings compared to many monocots. Some of these herbicides, such as DCPA and dithiopyr, are excellent turf herbicides, controlling annual grass growth and germination but not interfering with the growth of the perennial turf grasses.

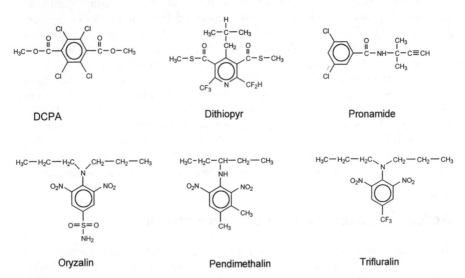

Fig. 9.1. Examples of chemical structures of some of the compounds classified as mitotic disrupter herbicides.

Although originally marketed as a herbicide, CIPC has been used effectively as a potato antisproutant. Perhaps the most unusual uses of these herbicides have been in the control of infectious human parasites (*Leishmania*) through oral administration of trifluralin (Chan et al. 1991; Benbow et al. 1998; Pfitzer et al. 1998). This is probably one of the few examples where a compound that has been developed as a herbicide has found a secondary use in medicine.

9.3
Mode of action

As mentioned above, one of the most characteristic effects of these herbicides is root-tip swelling into club-shaped structures. An identical morphology is induced by the classic mitotic disrupter alkaloid colchicine, and it was this similarity in effect that first gave clues as to the mode of action of the mitotic disrupter herbicides (Hess 1987; Vaughn and Vaughan 1988). Although rather high concentrations (more than 1 mM) are required to obtain root-tip swelling by colchicine (Vaughn and Vaughan 1988), micromolar or lower concentrations are all that are required to produce the same effect with a number of these herbicides.

The similarities between the effects of colchicine and the mitotic disrupter herbicides on the level of gross morphological effects, such as root tip clubbing, spurred investigations into the site of action of colchicine: the microtubule. Colchicine treatment of plant cells results in the loss of all microtubules and the appearance of characteristic C-mitoses in which the chromosomes proceed to a very condensed state of prometaphase but are unable to progress further in the mitotic cycle (Fig. 9.2a). The chromosome complement is present at the 4C-level and the nuclear membrane reforms around the chromosomes, resulting in a lobed 4C-nucleus. Cortical microtubules, which are involved in determining cell shape, are lost as well, resulting in a cell that expands isodiametrically rather than along the axis of elongation. The lack of proper cell expansion in the elongation zone and the mitotic arrests in the root meristem result in the club-shaped root morphology.

This same symptomatology is found after treatment with many of the mitotic disrupter herbicides as well, again indicating a commonality of the modes of action for the plant alkaloid colchicine and these herbicides. All microtubule arrays are affected by the mitotic disrupter herbicides, although the sensitivity of the arrays varies (Hoffman and Vaughn 1994). In general, the symptom found at the lowest concentrations of mitotic disrupter herbicides is on the phragmoplast microtubule array (Fig. 9.3). Irregular and disorganized phragmoplast arrays and, consequently, irregular and abnormal cell plates, are found (Fig. 9.3b,c). For some of the less effective herbicides, such as DCPA, this is virtually the only symptomatology noted, except in very sensitive species (Holmsen and Hess 1985; Vaughan and Vaughn 1990; Lehnen and Vaughn 1991a; Hoffman and Vaughn 1994).

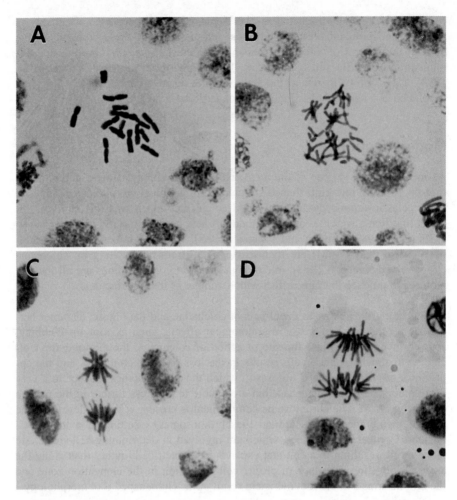

Fig. 9.2A-D. Root tip squashes of onion seedlings treated for 24 h with various mitotic disrupter herbicides stained with Feulgen reagent. **A** C-mitosis in which all the chromosomes are arranged in a "ski-pair" fashion after treatment with dithiopyr. **B** multipolar mitosis, where the chromosomes have moved to many poles is common after treatment with CIPC. **C,D** Examples of star anaphase configurations, after treatment with terbutol, with their arms splayed from these centres.

At increasing concentrations of herbicide, effects are noted on the spindle microtubule array (Fig. 9.4). These include multipolar mitosis (Fig. 9.2B, 9.4C,D), star anaphase (Fig. 9.2C,D, Fig. 9.4E), and fully arrested C-mitotic type figures (Fig. 9.2A). Multipolar mitosis (e.g. Lehnen and Vaughn 1992) seems to occur at lower concentrations than the C-mitotic figures, indicating that the multipolar divisions are a result of spindle microtubule formation being incompletely inhibited (Holmsen and Hess 1985; Hoffman and Vaughn 1994). The star anaphase figures are produced most regularly by the herbicide terbutol (Lehnen et al. 1992). In these figures, the chromosomes are gathered into star-like aggregations (Fig. 9.2C,D), centred on an accumulation of endomembrane which apparently serves as the microtubule-organizing center (MTOC) for this unusual array (Lehnen et al. 1992; Vaughn and Harper 1998). Star telophase figures (Fig. 9.3D) generally consist of microtubules that persist at the poles even though the phragmoplast array is already established and the phragmoplast forms around the star-like nucleus, resulting in an undulating rather than a smooth, plate-like structure (Lehnen et al. 1992).

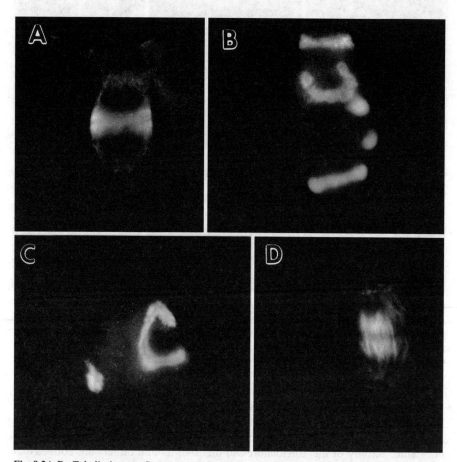

Fig. 9.3A-D. Tubulin immunofluorescence micrographs of phragmoplast microtubule arrays and the effects of various disrupters on these arrays. **A** Control phragmoplast of onion, in which the two ends of the phragmoplast are reaching towards the parental cell wall. **B** Oat cell treated with the herbicide DCPA where the phragmoplast contains extensive microtubular arrays that are misoriented in many different planes throughout the cell. **C** onion root-tip cell treated with CIPC with a prominent C-curved phragmoplast and a smaller phragmoplast fragment within the same cytoplasm. **D** Star telophase from a terbutol-treated onion root-tip cell. Although there is a prominent phragmoplast array, a number of microtubules are still arranged around the starburst of microtubules from the star-anaphase configuration.

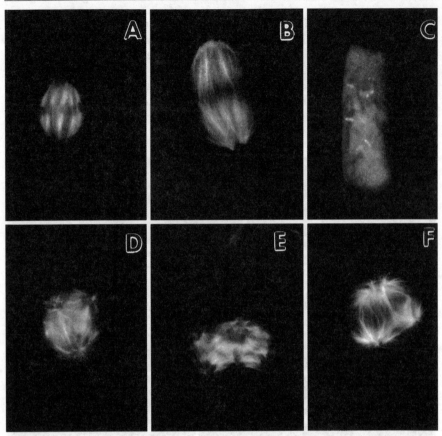

Fig. 9.4A-F. Tubulin immunofluorescence micrographs from metaphase (**A,C**) and anaphase (**B,D-F**) of onion root tips. **A** Control metaphase microtubule configuration. **B** Control anaphase configuration. **C** Kinetochore tufts of microtubules after dithiopyr treatment. **D** multipolar mitosis after CIPC treatment. **E** Multipolar mitosis with five different apparent centres of chromosome concentration after terbutol treatment. **F** Multi-star anaphase after terbutol treatment. The **upper pole** is a typical star anaphase configuration, whereas the **lower portion** of this cell contains two star centres.

Microtubules are found even less frequently at the highest concentrations of herbicides, generally with the only microtubules present in small kinetochore tufts (Fig. 9.4C). It is likely that these accumulations are the result of a microtubule-nucleating activity of the kinetochores so that free tubulin dimers are collected at the kinetochores. During treatment with the herbicides dithiopyr and pronamide, these kinetochore tufts are noted commonly, although all other microtubules are absent in these cells (Bartels and Hilton 1973; Vaughan and Vaughn 1987; Armbruster et al. 1991; Lehnen and Vaughn 1991b). At first, it was thought that these various abnormalities (phragmoplast disarray, multipolar mitoses, and kinetochore tufts) might be caused by several independent sites of action of these herbicides (Vaughn and Lehnen 1991). However, Hoffman and Vaughn (1994) showed that all the herbicides in this class could produce the same series of effects, but that the herbicides varied greatly in efficacy. Weak herbicides tend to produce just the phragmoplast or the multipolar mitosis effects (Holmsen and Hess 1985), whereas the most effective herbicides eliminate all microtubule arrays. By lowering the concentration of these very effective herbicides, the effects of the weak herbicides could be mimicked in a concentration-dependent manner, indicating a common mechanism of action (Hoffman and Vaughn 1994). Data from cross-resistance of herbicide-resistant mutants (see Chap. 8) helped to substantiate the congruence of sites of action (Vaughn et al. 1987; Smeda et al. 1992) as do data from molecular modelling (Ellis et al. 1994). For example, the green foxtail biotype that was selected for resistance as a trifluralin-resistant biotype is also resistant to terbutol and amiprophos methyl, although neither of these herbicides had been used at the locations where these resistant biotypes appeared (Smeda et al. 1992). Similarly, molecular modelling predicted that the phosphoric amides and dinitroaniline herbicides would occupy a similar binding site (Ellis et al. 1994) and dinitroaniline-resistant biotypes are indeed cross-resistant to phosphoric amide herbicides (Vaughn et al. 1987; Smeda et al. 1992).

In certain lower land plants, arrays associated with developing spermatogenous cells are posttranslationally modified by acetylation of the alpha tubulin (Hoffman and Vaughn 1995). These arrays are very stable, even in the presence of herbicide concentrations that induced complete loss of microtubules in all other arrays. Hoffman and Vaughn (1996) demonstrated that the arrays were sensitive to the herbicides during their formation but, once acetylated, the microtubules appeared resistant to the herbicide. Because acetylated arrays are turned over much more slowly than other microtubule arrays, the mechanism of these disrupters must be based upon the dynamics of microtubule assembly and disassembly.

Microtubules may be assembled in vitro from tubulin heterodimers and direct proof of the mode of action might be obtained from such in vitro studies. Oryzalin, pronamide, amiprophos methyl and CIPC have been directly examined for their effects on microtubule polymerization in the presence of the microtubule depolymerization blocker taxol (Akashi 1988; Morejohn and Fosket 1984; Morejohn et al. 1987). The levels of herbicides required to cause substantial inhibition of microtubule polymerization are greater than would be expected from the effects of these herbicides observed in vivo. Taxol promotes microtubule stability and thus might compete with a herbicide-caused inhibition of dimer addition to the microtubules. This is supported by experiments where plants were pretreated with taxol and where microtubules in the treated cells were found to be resistant to the disrupter herbicides (Vaughn and Vaughan 1991). Polymerization of plant microtubules in the presence of dimethyl sulfoxide is also possible and generally results in a more complete inhibition of microtubule assembly by these herbicides in vitro (Vaughn and Vaughan 1990; K.C. Vaughn, unpubl. results). However, tubulin isolated from a resistant biotype of goosegrass (*Eleusine indica*) was able to assemble microtubules *in vitro* even in the presence of these herbicides (Vaughn and Vaughan 1990). These data were the first biochemical indication that the differences in the dinitroaniline-resistant goosegrass biotypes were, in fact, caused by alterations at the level of microtubule dynamics.

One might expect a substochiometric relationship between the binding of the herbicide and its ability to inhibit microtubule polymerization. Based upon models of microtubule assembly inhibitors such as colchicine, one might expect the disrupters to react with free tubulin heterodimers in the cytoplasm. The tubulin heterodimer / herbicide complex, by binding to the growing (+) end of the microtubule, would prevent the further addition of microtubule subunits. Because the microtubule is dynamic, one would still expect the loss of heterodimers to continue at the (-) end of the microtubule. Thus, by the addition of the herbicide-heterodimer complex at the growing end of the microtubule, the herbicide could effectively inhibit microtubule polymerization far below the level of equal molar concentrations of herbicide and tubulin heterodimer. Binding ratios have been found that are either equimolar or approximatively 4:1, depending upon species and tubulin preparation (Hess and Bayer 1977; Strachen and Hess 1983; Morejohn and Fosket 1984). Tubulin is extremely sensitive to proteolysis and therefore the variability might be related to the degree of proteolysis that has occurred during isolation and/or binding assays. A protein of around 68 kDa has been shown to bind to dithiopyr (Molin et al. 1988), which is a molecular weight in the range of a number of microtubule-associated proteins, but well above that of tubulin monomers.

The lipophilic nature of the majority of the mitotic disrupter herbicides and the sites at which plant cells nucleate microtubules may also favour the substochiometric inhibition of microtubule assembly. Molin and Khan (1997) recently reviewed the interactions of lipids and membranes with the dinitroaniline herbicides. Addition of these herbicides to plant tissues results in the rapid accumulation of the herbicide from the surrounding medium into the plant tissue. This can often be demonstrated visually as the orange solutions of dinitroaniline herbicides lose their colour, whereas the incubated root tissue becomes distinctly orange. Recent data indicate that microtubules in higher plants are nucleated at the nuclear envelope and/or the endomembrane system (Vaughn and Harper 1998). Thus, these lipophilic herbicides might accumulate in these subcellular sites and reach much higher effective concentrations, allowing for a more efficient inhibition of microtubule assembly.

An alternative theory on the action of mitotic disrupter herbicides was proposed by Hertel et al. (1980), in which changes in cellular calcium concentrations are altered by the herbicide. Because microtubule assembly is very sensitive to changes in calcium concentration, Hertel et al. (1980) proposed that leakage of calcium from mitochondria would cause changes in cytosolic calcium that are sufficient to cause changes in microtubule assembly or stability. Although many of these herbicides did cause changes in mitochondrial calcium leakage, 10-100 times higher concentrations were required to see this effect if compared to the herbicide concentrations that inhibit microtubule assembly in vitro or in vivo

(Hoffman and Vaughn 1994). Moreover, although oryzalin and trifluralin could inhibit microtubule polymerization only in the sensitive biotype of goosegrass, but not in the resistant biotype, calcium loss from mitochondria occurred in both biotypes (Vaughn and Vaughan 1991). Thus, although these effects on calcium distribution may be real, they probably contribute little to the effects on microtubules *in vivo*.

9.4
Metabolism and natural tolerance mechanisms

Most studies of dinitroaniline herbicides indicate that very little metabolism of these herbicides occurs via plant-derived mechanisms, although a number of microbial metabolites are known. An investigation of the metabolites of trifluralin indicates that all but one of the metabolites are herbicidally inactive, and even that one was much less effective than the parent molecule (Vaughn and Koskinen 1987). An exception to the lack of detoxification by herbicide metabolism is the herbicide pendimethalin, which may be metabolized via a P450-type oxidase to an inactive compound, as recently demonstrated in a biotype of black-grass from the United Kingdom (James et al. 1995). This biotype displays resistance to pendimethalin but not to other dinitroanilines and is cross-resistant to other herbicides that are metabolized via P450 systems. The resistance in the multiply resistant ryegrass biotypes in Australia, however, may be due to mechanisms of resistance different from P450-mediated oxidation, as resistance was observed to herbicides that are not metabolized by the P450 system (McAlister et al. 1995). Potentially, such mechanisms of herbicide resistance through P450 oxidases could be introduced into crop plants that normally are sensitive to pendimethalin.

Hilton and Christianson (1972) were the first to note a correlation between the amount of lipid in the seed and the level of tolerance to dinitroaniline herbicides. Lipid bodies, found in the cotyledons, would sequester the lipophilic herbicide away from the site of action, the microtubule. Pretreatment of seeds of plants with normal sensitivity with lipid solutions could also protect young seedlings from the effects of these herbicides. The additional lipid would sequester the herbicide and the sensitive plant could be rendered herbicide-tolerant. In biotypes of cotton with lysigenous (gossypol-containing) glands, similar kinds of resistance mechanisms exist. The lipid-rich gossypol gland is a site of herbicide sequestration (Strang and Rogers 1972) and glandless lines of cotton are much more sensitive to the effects of several lipophilic herbicides, including the unrelated lipophilic herbicide norflurazon, a carotenoid synthesis inhibitor (Stegink and Vaughn 1988).

One puzzling note on the natural tolerance of plants to mitotic disrupter herbicides concerns members of the carrot family. Although dicots are, in general, less sensitive to dinitroanilines than monocots, the carrot family is exceptional in their nearly complete resistance to these compounds (Vaughan and Vaughn 1988). The carrots are not especially large seeded nor do they contain the copious amounts of lipids or special sequestration sites found in other very tolerant crops or weeds. Thus, it is unclear why they should be so insensitive to this whole group of compounds. Studies are presently under way to determine potential changes in tubulin proteins in these plants to determine whether there might be any correlation between the high natural herbicide resistance and alterations in tubulin proteins (D. Prochaska and K.G. Wilson, pers. Comm.).

Among crop plants, there is some variation in the sensitivity of given cultivars to these herbicides. This is an area where more attention of plant breeders to variation in herbicide tolerance could allow for registration of a herbicide for a new crop species. Variation among the Cucurbitaceae species to dinitroaniline herbicides is at present the only example where this phenomenon has been investigated genetically (Adeniji and Coyne 1981; Poe and Coyne 1988). Inducing mutants with ethyl methane sulfonate was totally unsuccessful in isolated *Arabidopsis* mutants that were resistant to dinitroaniline herbicides, despite the screening of millions of M2 *Arabidopsis* seedlings (K.C. Vaughn, unpubl.; for other herbicides: D.P. Snustad, unpubl.). Similar selection schemes have been successful in other species (e.g. Heim et al. 1989). It is not known whether the modifications of tubulin responsible for dinitroaniline resistance are less mutatable than other loci in *Arabidopsis* or whether these mutations are lethal in this species. The occurrence of dinitroaniline-resistant biotypes in both monocots and dicots indicates that such mutations in other genera are possible and similar selection schemes in crop plants may result in dinitroaniline-resistant lines. Indeed, tissue culture selection schemes using tobacco protoplasts resulted in the selection of mutants resistant to mitotic disrupter herbicides (e.g. Blume and Strashnyuk 1998).

9.5
Induction of tetraploidy

As already mentioned, the herbicides in this group act very much like antimicrotubule drugs such as colchicine or vinblastine, but are much more effective, on a molar basis, in disrupting mitosis in plants (Vaughn and Vaughan 1988). One of the uses for these antimicrotubule drugs has been the induction of tetraploidy in normally diploid species. For this process, relatively high concentrations (up to 2-3% active ingredient) are used in the treatment, and a relatively low success rate is obtained for most species. In general, the plants are treated over brief periods of time with a colchicine solution; this treatment arrests mitosis at prometaphase such that the cells have four sets of chromosomes. This colchicine is then washed out, which allows the plants to recover, but the cells now maintain the 4C content of DNA. Because the mitotic disrupter herbicides work in the same way as colchicine does, attempts have been made to use them as more efficient and less toxic (with respect to mammalians) alternatives to induce tetraploidy in plants. A number of successful attempts have been made, including the doubling of anther-derived haploid maize cultures to the diploid level (Wan et al. 1991) and beet ovule cultures (Hansen and Anderson 1996) with oryzalin. Seedlings of a number of different ornamental species have been rendered tetraploid using either oryzalin or trifluralin, including species that have proven recalcitrant to DNA duplication during colchicine treatment (K.C. Vaughn, unpubl. results). Thus, it is likely that these herbicides will see more use in the future as agents in inducing tetraploidy more efficiently than colchicine and with much lower risks for the users. Certainly, the backyard plant breeder, without access to the laboratory conditions for handling colchicine, might be better served to use these alternatives for generating tetraploid plants.

9.6
Conspectus

Mitotic disrupter herbicides have been utilized much beyond their herbicidal use to investigate the basic biology of plant microtubules. Here, they have been useful in determining the relative stability of microtubule arrays and the cellular consequences of microtubule loss. Experiments in which the herbicides have eliminated the microtubules and the microtubules were then allowed to regrow have allowed the identification of subcellular sites responsible for microtubule nucleation and organization (Vaughn and Harper 1998). The use of these compounds as selective agents for plant transformation or in traditional breeding schemes in which the most tolerant individuals are selected (e.g. Adeniji and Coyne 1981) should allow the more widespread use of these compounds. This may result in more effective weed control in cropping conditions where these herbicides are not an option presently. Similarly, the more frequent use of these herbicides for the induction of tetraploidy could markedly improve the ability of plant breeders to improve the quality of both ornamental and agronomic species.

Acknowledgements. Research on the mitotic disrupter herbicides in my laboratory has been supported by several in-house postdoctoral positions and an USDA-NRI Competitive Grant to support postdoctoral research associates. These include: Martin A. Vaughan, Larry P. Lehnen, John C. Hoffman, Reid J. Smeda and Neil A. Durso. Helpful discussions have been carried out with Drs. Reiner Kloth and William T. Molin. Technicians Ruth Jones and Lynn Libous-Bailey have provided tremendous technical assistance in performing the numerous microscopic studies required in these experiments.

References

Adeniji AA, Coyne DP (1981) Inheritance of resistance to trifluralin toxicity in *Cucurbita moschata* Poir. Hort Sci 16: 774-775

Akashi T (1988) Effects of propyzamide on tobacco cell microtubules in vivo and in vitro. Plant Cell Physiol 29: 1053-1062

Armbruster BL, Molin WT, Bugg MW (1991) Effects of the herbicide dithiopyr on cell division in wheat root tips. Pestic Biochem Physiol 39: 110-120

Bartels PG, Hilton JL (1973) Comparison of trifluralin, oryzalin, pronamide, propham and colchicine treatment on microtubules. Pestic Biochem Physiol 3: 462-472

Benbow JW, Bernberg EL, Korda A, Mead JR (1998) Synthesis and evaluation of dinitroanilines for treatment of cryptosporidiosis. Antimicrob Agents Chemother 42: 339-343

Blume YB, Strashnyuk NM (1998) Alterations of beta-tubulin in *Nicotiana plumbaginifolia* confers resistance to amiprophos methyl. Theor Appl Genet 97: 464-472

Chan MM, Triemer RE, Fong D (1991) Effects of the anti-microtubule drug oryzalin on growth and differentiation of the parasite *Leishmania mexicana*. Differentiation 46: 15-21

Ellis JR, Taylor R, Hussey PJ (1994) Molecular modeling indicates that two chemically distinct classes of anti-mitotic herbicide bind the same receptor site(s). Plant Physiol 105: 15-18

Hansen NJP, Anderson SB (1996) In vitro chromosome doubling potential of colchicine, oryzalin, trifluralin and APM in *Brassica napus* microspore culture. Euphytica 88: 159-164

Heim DR, Roberts JI, Pike PD, Larrinua IM (1989) Mutation of a locus of *Arabidopsis thaliana* confers resistance to the herbicide isoxaben. Plant Physiol 90: 146-150

Hertel C, Quader H, Robinson DG, Marme D (1980) Anti-microtubular herbicides and fungicides affect Ca transport in plant mitochondria. Planta 149: 336-340

Hess FD (1987) Herbicide effects on the cell cycle of meristematic plant cells. Rev Weed Sci 3: 183-203

Hess FD, Bayer DE (1977) Binding of the herbicide trifluralin to *Chlamydomonas* flagellar tubulin. J Cell Sci 24: 351-360

Hilton JL, Christiansen MN (1972) Lipid contribution to selective action of trifluralin. Weed Sci 20: 290-294

Hoffman JC, Vaughn KC (1994) Mitotic disrupter herbicides act by a single mechanism but vary in efficacy. Protoplasma 179: 16-25

Hoffman JC, Vaughn KC (1995) Post-translational tubulin modifications in spermatogenous cells of the pteridophyte *Ceratopteris richardii*. Protoplasma 186: 169-182

Hoffman JC, Vaughn KC (1996) Spline and flagellar microtubules are resistant to mitotic disrupter herbicides. Protoplasma 192: 57-69

Holmsen JD, Hess FD (1985) Comparison of the disruption of mitosis and cell plate formation in oat roots by DCPA, colchicine and propham. J Exp Bot 36: 1504-1513

James EH, Kemp MS, Moss SR (1995) Phytotoxicity of trifluoromethyl- and methyl substituted dinitroaniline herbicides on resistant and susceptible populations of black-grass (*Alopecurus myosuroides*). Pestic Sci 43: 273-277

Lehnen LP, Vaughn KC (1991a) Immunofluorescence and electron microscopic investigations of DCPA-treated oat roots. Pestic Biochem Physiol 40: 47-57

Lehnen LP, Vaughn KC (1991b) Immunofluorescence and electron microscopic investigations of the effects of dithiopyr on onion root tips. Pestic Biochem Physiol 40: 58-67

Lehnen LP, Vaughn KC (1992) The herbicide sindone B disrupts spindle microtubule organizing centers. Pestic Biochem Physiol 44: 50-59

Lehnen LP, Vaughan MA, Vaughn KC (1992) Terbutol affects spindle microtubule organizing centers. J Exp Bot 41: 537-546

McAlister FM, Holtum JAM, Powles SB (1995) Dinitroaniline herbicide resistance in rigid ryegrass (*Lolium rigidum*). Weed Sci 43: 55-62

Molin WT, Khan RA (1997) Mitotic disrupter herbicides: recent advances and opportunities. In: Roe RM (ed) Herbicide activity: toxicology, biochemistry and molecular biology. IOS Press, Burke, VA, USA, pp 143-158

Molin WT, Lee TC, Bugg MW (1988) Purification of a protein which binds to MON 7200. Plant Physiol (Suppl) 86: 21

Morejohn LC, Fosket DE (1984) Inhibition of plant microtubule polymerization in vitro by the phosphoric amide herbicide amiprophos methyl. Science 224: 874-876

Morejohn LC, Bureau TC, Molé-Bajer J, Bajer AS, Fosket DE (1987) Oryzalin, a dinitroaniline herbicide, binds to plant tubulin and inhibits microtubule polymerization in vitro. Planta 172: 252-264

Pitzer KK, Werbovetz N, Brendke JJ, Scovill JP (1998) Synthesis and biological evaluation of 4-chloro-3,5-dinitrobenzotrifluoride analogues as antileishmanial agents. J Med Chem 41: 4885-4889

Poe RR, Coyne DP (1988) Differential *Cucurbita* tolerance to the herbicide trifluralin. J Am Soc Hort Sci 113: 35-40

Sabba R, Vaughn KC (2000) Herbicides that inhibit cellulose biosynthesis. Weed Sci (in press)

Smeda RJ, Vaughn KC, Morrison IN (1992) A novel pattern of herbicide cross-resistance in a trifluralin-resistant biotype of green foxtail [*Setaria viridis* (L.) Beauv.] Pestic Biochem Physiol 42: 227-241

Steginck SJ, Vaughn KC (1988) Norflurazon (SAN-9789) reduces abscisic acid levels in cotton seedlings: a glandless isoline is more sensitive than its glanded counterpart. Pestic Biochem Physiol 31: 269-275

Strachen SD, Hess FD (1983) The biochemical mechanism of the dinitroaniline herbicide oryzalin. Pestic Biochem Physiol 20: 141-150

Strang RH, Rogers RL (1972) A microautoradiographic study of C-14 trifluralin absorption. Weed Sci 19: 363-369

Vaughan MA, Vaughn KC (1987) Pronamide disrupts mitosis in a unique manner. Pestic Biochem Physiol 28: 182-193

Vaughan MA, Vaughn KC (1988) Carrot microtubules are dinitroaniline resistant. I. Cytological and cross-resistance studies. Weed Res 2: 73-83

Vaughan MA, Vaughn KC (1990) DCPA causes cell plate disruption in wheat roots. Ann Bot 65: 379-388

Vaughn KC, Harper JDI (1998) Microtubule organizing centers and nucleating sites in land plants. Int Rev Cytol 181: 75-149

Vaughn KC, Koskinen WC (1987) Effects of trifluraline metabolites on goosegrass (*Eleusine indica*) root meristems. Weed Sci 35: 36-44

Vaughn KC, Lehnen LP (1991) Mitotic disrupter herbicides. Weed Sci 39: 450-457

Vaughn KC, Vaughan MA (1988) Mitotic disrupters from plant cells, effects on plants. In: Cutler HG (ed) Biologically active natural products: potential uses in agriculture. American Chemical Society, Washington DC, pp 273-293

Vaughn KC, Vaughan MA (1990) Structural and biochemical characterization of dinitroaniline-resistant *Eleusine*. In: Green MB, Le Baron HM, Moberg WK (eds) Managing resistance to agrochemicals: from fundamental research to practical strategies. Am Chem Soc, Washington DC, pp 364-375

Vaughn KC, Vaughan MA (1991) Dinitroaniline resistance in *Eleusine indica* may be due to hyperstabilized microtubules. In: Caseley JC, Cussans GW, Atkins RK (eds) Herbicide resistance in weeds and Crops. Butterworth-Heinemann, Oxford, pp 177-186

Vaughn KC, Marks MD, Weeks DP (1987) A dinitroaniline resistant mutant of *Eleusine indica* exhibits cross-resistance and supersensitivity to antimicrotubule herbicides and drugs. Plant Physiol 83: 956-964

Wan Y, Duncan DR, Rayburn AI, Petolino JF, Widholm JM (1991) The use of antimicrotubule herbicides for the production of doubled haploid plants from anther-derived maize callus. Theor Appl Genet 81: 205-211

Subject Index

Figures are indicated by italics.

Printing: Weihert-Druck GmbH, Darmstadt
Binding: Buchbinderei Schäffer, Grünstadt